DIGITAL INNOVATION

INNOVATION PROCESSES IN VIRTUAL CLUSTERS AND DIGITAL REGIONS

Series on Technology Management

Series Editor: J. Tidd (Univ. of Sussex, UK)

Published

Vol. 1 Engines of Prosperity
Templates for the Information Age
by G. R. Ungson *(Univ. of Oregon, USA)* & J. D. Trudel *(The Trudel Group, USA)*

Vol. 2 The Knowledge Enterprise
Implementation of Intelligent Business Strategies
edited by J. Friso den Hertog *(MERIT, Maastricht University and Altuïtion bv, 's Hertogenbosch, The Netherlands)* &
E. Huizenga *(Altuïtion bv, 's Hertogenbosch, The Netherlands)*

Vol. 3 From Knowledge Management to Strategic Competence
edited by J. Tidd *(Univ. of Sussex, UK)*

Vol. 4 Japanese Cost Management
edited by Y. Monden *(Univ. of Tsukuba, Japan)*

Vol. 5 R&D Strategy on Organisation
Managing Technical Change in Dynamic Contexts
by V. Chiesa *(Univ. degli Studi di Milano, Italy)*

Vol. 6 Social Interaction and Organisational Change
Aston Perspectives on Innovation Networks
edited by O. Jones *(Aston Univ., UK)*, S. Conway *(Aston Univ., UK)* & F. Steward *(Aston Univ., UK)*

Vol. 7 Innovation Management in the Knowledge Economy
edited by B. Dankbaar *(Univ. of Nijmegen, The Netherlands)*

Vol. 8 Digital Innovation
Innovation Processes in Virtual Clusters and Digital Regions
edited by G. Passiante (*Univ. of Lecce, Italy*), V. Elia (*Univ. of Lecce, Italy*) & T. Massari (*Univ. of Lecce, Italy*)

Forthcoming

Vol. 9 Service Innovation
Organisational Responses to Technological Opportunities and Market Imperatives
edited by J. Tidd (*Univ. of Sussex, UK*) & F. M. Hull (*Fordham Univ., USA*)

SERIES ON TECHNOLOGY MANAGEMENT – VOL. 8

DIGITAL INNOVATION

INNOVATION PROCESSES IN VIRTUAL CLUSTERS AND DIGITAL REGIONS

Editors

Giuseppina Passiante
Valerio Elia
Tommaso Massari

University of Lecce, Italy

ICP Imperial College Press

Published by

Imperial College Press
57 Shelton Street
Covent Garden
London WC2H 9HE

Distributed by

World Scientific Publishing Co. Pte. Ltd.
5 Toh Tuck Link, Singapore 596224
USA office: Suite 202, 1060 Main Street, River Edge, NJ 07661
UK office: 57 Shelton Street, Covent Garden, London WC2H 9HE

British Library Cataloguing-in-Publication Data
A catalogue record for this book is available from the British Library.

DIGITAL INNOVATION: INNOVATION PROCESSES IN VIRTUAL CLUSTERS AND DIGITAL REGIONS

ISBN 1-86094-352-7

Printed in Singapore by World Scientific Printers (S) Pte Ltd

Contents

Introduction to Digital Innovation

Aldo Romano

Innovation is the key driver of economic activity and business success. Knowledge and learning are the main sources of value created by innovation.

The tremendous growth of IT and digital networks are enabling and accelerating the dissemination of "codified knowledge" and the creation of new knowledge. The shift from analog to digital technologies, from multi-level hierarchical computing architectures to new internetworked models are enabling the commodification of knowledge and the pervasiveness of innovation generated by knowledge diffusion and creation. Innovation is ubiquitous: in every firm, in every industry, in every geography.

Transcending both spatial and temporal boundaries, digital technologies and IT networks are creating an entirely different competitive space and social experience. Therefore, we interpret digital innovation as innovation enabled by digital technologies. In the terminology of Schumpeterian economic analysis, digital innovation could be considered as the emergent "creative destruction process" boosted by enabling ICTs related to microelectronics. Particularly, two forces drive the potential of digital innovation: the explosion of digital connectivity and the rapid emergence of universal technical standards for communication that allow everybody to communicate with everyone else at essentially zero cost (Evans and Wurster, 2000). These forces contribute to generating the critical features of digital networks: openness, interoperability, scalability and extensibility. These features are understood by economists generalising the concept of the Internet's bearer service through the idea that the Internet acts as general purpose technology or platform technology (McKnight, Valer, and Katz, 2001). Digital innovation is fuelled by this platform technology. The virtuous fertilization between the old Economy and Internet worked e-technologies represents the overall impact of digital innovation. From this fertilization emerges a new economic landscape which we call the Knowledge-based Network Economy (KBNE).

Recent literature is focused on and debating the new attributes and rules emerging in the KBNE. A fundamental attribute of the KBNE is "Virtualness Perspective" as a strategic asset of overall organisation belonging to market and non-market contexts. "Virtualness Perspective" adds to the economic and social benefits arising from geographical proximity, with the advantages coming from organisational proximity based on the ramification of new ICTs. According to Teece, digital innovation supported by advanced communication infrastructure can facilitate entrepreneurialism by stripping out barriers to communication. As Teece states, "It challenges existing organization boundaries, divisions, and hierarchies and permits formal organization to be more specialized and responsive. Interorganizationally, networked organizations have blurred and shifting boundaries and they function in conjunction with other organizations" (2000).

Particularly the "Virtualness Perspective" reinforces the effects of the laws of increasing returns. This law is considered as a new rule of the Knowledge-based Economy, that contrasts with the traditional concept of "diminishing returns" based on the notion of scarcity of resources (Kevin Kelly, 1998). Kelly provides a definition of the law of increasing returns: "Value explodes with membership, and the value explosion sucks in more members compounding the result". Kelly also sees this effect in the way areas such as Silicon Valley grow. "Each successful start-up attracts other start-ups, which in turn attract more capital and skills and yet more start-ups. (Silicon Valley and other high tech industrial regions are themselves tightly coupled networks of talent, resources, and opportunities). This is because increasing returns are based on "mechanisms of positive feedback that operate — within markets, business and industries — to reinforce that which gains success or aggravate that which suffers loss". (Arthur, 1996).

Therefore the "Virtualness Perspective Effect" boosts the competitive situation related to increasing returns. Indeed a ubiquitous network allows all members to share data, information, knowledge and to spread technology costs. According to Metcalfe's law, the value of a network increases with the square of the number of its participants. When communication is valuable, a new user creates value for everyone on the network who wants to communicate. The "Virtualness Perspective" thus enables and accelerates the shift towards flexible production systems and inter-organisational coevolution. Kevin Kelly offers a fine representation of this shift. An automobile maker in the industrial age maintains control over all aspects of the car's parts and construction. An automobile maker in the Network Economy will establish a web of standards and outsourced suppliers, encouraging the web itself to invent the car, seeding the system with the knowledge it gives away, engaging as many participants as broadly as possible, in order to create a virtuous loop where every member's success is shared and leveraged by all.

The "Virtualness Perspective" enables and accelerates the shift from an economic organisation model based on "command and control" hierarchies to an economic organisation model which John H. Dumming calls "flexible capitalism". Peter Schwarz, Eamonn Kelly and Nicole Boyer metaphorically call these two models "citadel organisations" and "web organisations" respectively. The "citadel organisation" model was ideal for industrial based activities where cost-efficiencies, increased productivity and economies of scale were of paramount importance. In this model, the future was relatively certain and predictable. The "web organisations" models spring from the contrary belief that the world is increasingly complex, full of rapid discontinuous change, and unpredictable. The challenge today is managing the "interface" between these two very different worlds (P. Schwartz, E. Kelly, N. Boyer, 1999). The response to this challenge requires rethinking, redirecting, and re-positioning the traditional theoretical and practical cultural frameworks, related to different fundamental issues.

In this book, some issues are discussed and analyzed in order to highlight the new paradigm emerging from the competitive space of connections, relationships and patterns of interaction. Particular focus is devoted to the impact of digital innovation, including the "Virtualness Perspective", on the inter-relationship surrounding economic organisation and spatial change. The main questions discussed and analysed concern the impact of digital innovation on the evolution of industrial clusters and on regional development dynamics.

The system of innovation approach is followed by the authors in their analysis and discussion, with a particular focus on a wide range of literature regarding:

- Innovation as competitiveness-enhancing interactive learning processes.
- Clustering, networking, regional strategy-setting processes as the main elements in an efficient context for interactive learning.

An invariant is present in the phenomena analysed and interpreted by the authors: digital innovation highlights the relevance of local, regional, and national comparative advantages based on intellectual capital and technological infrastructure, where intellectual capital is framed in three components: human capital, structural capital and social capital. Intellectual capital is also interpreted as a collective capability to initiate and sustain significant changes, to work effectively with forces that shape changes. This collective capability is defined as "Knowledge Leadership", where knowledge is viewed as a "capacity for effective action and innovation". Knowledge Leadership is more about "competing on-the-edge strategies" than strategising on clear industry or geographic boundaries, predictable competition or a knowable future. Knowledge Leadership requires a mental shift from forecasting approaches to the ability to foresee.

Classic forecasting is based on a deterministic model. Foresight, on the other hand, is the ability to see what is emerging, to influence the future by responding to and influencing it when it is emerging. With foresight we can see changes coming and respond before a crisis arises. We have an opportunity to focus on resources in ways that will begin to influence the future immediately (Sanders, 1998). From this shift emerges the relevance of strategic thinking for Knowledge Leadership. Indeed, the purpose of strategic thinking is to help an organisation to identify, respond to, and influence changes in its environment (Sanders, 1998). It is a search for information, knowledge and options which will ensure an ongoing competitive advantage for the organisation, local and regional community, given their core competencies, strengths and experiences. This profile of Knowledge Leadership highlights the complexity of the landscape influenced by digital innovation. This landscape is configurable as space of relationship connections and patterns of interaction.

In this landscape organisations and local communities seek to become a Complex Adaptive System (CAS): non-linear evolutionary systems that are constantly promising and incorporating new information and new knowledge. "In CAS, change is constant because of the flow of new information, but dramatic change occurs when something tips the balance. If the system is sensitive to new information, it goes through a period of adaptation of which a new pattern or shape emerges. And, sometimes the changes are dramatic enough to be recognised as transformative. The system before the change is vastly different from the system after the change" (Sanders, 1998). Therefore "knowledge assets" management and creation as well as Knowledge Leadership are the drivers of success in the KBNE.

Another invariant arising from the authors' discussion concerns the policy implications related to the KBNE landscape. Finding the right institutional arrangements to govern an emerging KBNE is creating many problems for governments. In an industrial economy, the government is concerned about the allocation of scarce resources whereas in the KBNE the goal will be to foster knowledge creation and the discovery of new things (P. Schwartz, E. Kelly, N. Boyer, 1999). In the KBNE, governments will be more focused on creating "frameworks" policies that set the context or "road-map" from which knowledge activities happen (Stevens, 1996). This approach is more nurturing and stimulates drivers of growth such as technological innovation, and is open to new ideas and outside influences. "Like corporations, bureaucracies will have to become nimble learning organisations". The character and style of policy-making may come to resemble "action-oriented" research, which tests several directions or pilot projects at the same time and then makes modifications and directional decisions through a series of learning loops. "This is a more grass-roots, responsive, and experimental way to govern" (P. Schwartz, E. Kelly, N. Boyer, 1999).

The central challenge for governments is how to create the right education for the KBNE while improving the access for more people. Particularly, the challenge concerns "business schools" which are traditionally focused on education programs for managers and leaders. This challenge calls for radical innovation both in knowledge organisation and in learning strategies, in order to design learning contexts suitable for enhancing the collective intelligence and capacity to create. This radical innovation should be related to some fundamental issues:

- Searching for a new Science of Business in terms of new conceptual and analytic tools for understanding and shaping adroit business response and for increasing the collective ability of an organisation to survive and prosper during the seemingly chaotic process of change. This requires scanning across different disciplines and industries as the key to spot emerging conditions, paradigm shifts and opportunities from the competitive landscape of the KBNE. This new openness creates opportunities for synergistic thinking and cooperative research, focused on the impact of digital innovation, on the rise of knowledge as a key economic resource, on the changing face of competition, on global market dynamics, and on government-driven changes.

- Searching for an integration of the various frameworks, in particular for linking strategic management and entrepreneurship theories of value creation. This calls for looking at the whole system and not just at their parts; the whole is a constantly changing field of connections, relationships and patterns of interactions. Innovation occurs as the result of new influences that disrupt existing patterns; out of these disruptions, new patterns emerge, seen as issues, trends and technological developments. Non-linear thinking is critical to recognise these patterns, because they are the result of relationships between orderly and disorderly states. This new strategic entrepreneurship paradigm requires a shift in the management paradigm from a focus on economic performance indicators (market share, sales, profit, ROI, cash flow) towards additional attention to process measures (time to market, process reliability, cost of quality) and towards reflective measures, related to knowledge management (people competencies, IT systems and databases, learning capabilities, management capabilities, intellectual property, and management philosophy).
- Redesigning the educational environment in terms of a learning organisation trying to gain increasing returns value from its network of academic and industrial partners. Creating Knowledge Leadership requires "constructed microcosms of real settings", in which teams of talented people can learn how to learn together, enhancing their collective intelligence and capacity to create. The process of these "microcosms", which Senge calls "learning laboratories", is a continual movement between practice and performance; it combines meaningful business issues with meaningful interpersonal dynamics.
- Activating learning strategies and learning processes based on knowledge management processes related to knowledge-creating company paradigms. The "microcosms of real settings" stimulate and sustain a combination of organisational adaptive and generative

learning processes. Adaptive learning is focused on disseminating and making accessible knowledge that already exists, and contributes to nurturing continuous improvement and problem solving processes. Generative learning is focused on creating new knowledge and commercializing it as a valuable product, service and/or new business. Generative learning, unlike adaptive learning, requires new ways of looking at the world; it is related to the creative tension, in the terms proposed by Senge.

Dealing with these issues means realising the favorable context to generate intellectual capital and Knowledge Leadership which are strategic assets in the environment driven by Digital Innovation.

Structure of the book

The book is made up of five chapters:

- Chapter 1: "Industrial clusters in the Net-Economy: empirical evidence and some theoretical approaches" by Giuseppina Passiante.
- Chapter 2: "A strategy to sustain the development processes of a 'smart community' in the Digital Economy" by Valerio Elia.
- Chapter 3: "A knowledge hub for regional development" by Mario Marinazzo.
- Chapter 4: "Net-Economy business models for SMEs operating in fragmented markets" by Tommaso Massari.
- Chapter 5: "Towards digital innovation leadership" by Giuseppina Passiante and Valerio Elia.

References

Evans P.B. and Wurster T.S. (2000) *Blown to Bits: How the New Economics of Information Transforms Strategy*. Harvard Business Scholl Press, Boston, Massachussetts.

McKnight L.W., Vaaler P.M. and Katz R.L. (2001) Introduction to creative destruction, in McKnight L.W., Vaaler P.M. and Katz R.L. (eds.), *Creative Destruction*. The MIT Press, Massachussetts.

Teece D.J. (2000) *Managing Intellectual Capital*. Oxford University Press, Oxford.

Kelly K. (1998) *New Rules for the New Economy: 10 Radical Strategies for a Connected World*. Penguin Books, London.

Arthur B. (1996) Increasing returns and the new world of business, *Harvard Business Review*. July–August, 100–109.

Schwartz P., Kelly E. and Boyer N. (1999) The emerging global knowledge economy, in *The Future of the Global Economy: Towards a Long Boom?* OECD, pp. 77–113.

Sanders T.I. (1998) *Strategic Thinking and the New Science: Planning in the Midst of Chaos, Complexity and Change*. Simons & Schuster, New York.

Stevens (1996).

Chapter Summaries

Chapter 1

Industrial Clusters in the Net-Economy: Empirical Evidence and Some Theoretical Approaches

In Chapter 1, Giuseppina Passiante organises the shift from geographical to virtual clusters within the more general issue of the impact of digital innovation on the interrelationships between the economic organisation and the spatial change of an economy.

More specifically, she focuses on the new forms of spatial organisation of the industrial clusters that are emerging in the Internet-based Economy. In these "virtual clusterisation forms", each enterprise adds one or more distinct aspects of product/service value to the value of the network, by exchanging digital knowledge with other members. A virtual cluster (VC) is considered the result of an integration process of different "core competencies" owned by each single partner, supported by the necessity to face the risks, costs and complexity of innovation. The learning processes that each member develops tend to develop around new sources of knowledge, instead of geographical or spatial criteria, generating new collective learning processes, allowed by the Internet-based technologies. In this view, the author emphasises the positive interference between organisational proximity and geographical proximity, as generated by these new organisational forms, characterised by ubiquitous provision and distribution of information in networks of agents. "Virtualness" is considered to be a relevant source of competitiveness that accelerates the pervasiveness of innovation — digital innovation — transcending local, regional and national boundaries.

As a new unit of analysis useful to describe these innovation processes within VCs, the author introduces the concept of Virtual Innovation System (VIS). The VIS concept is also presented as a useful theoretical framework for comprehending the meaning of the shift from the economy as an allocative mechanism to the economy as a learning process. Indeed, VIS dynamics is characterised by learning mechanisms and knowledge accumulation processes that are the key issues of the learning processes that develop in digital innovation processes. Particularly, VIS dynamics is interpreted as:

- An adaptive and cumulative learning process.
- A system of interactions among organisationally proximate agents and groups.
- A network logic.

In this view, VIS dynamics is described in terms of complex adaptive systems (CAS) behaviours, with an internal dynamic of self-sustaining cycles of knowledge exchanges between their organisations. In other terms, a VIS is characterised by a positive self-sustaining knowledge mechanism among its organisations, which develop highly interconnected webs of relations among them, generating a degree of "complexity" within the network.

Based on these VIS characteristics, some final policy implications are also presented, concerning the role of local/regional/national governments in fostering dynamic agglomeration economies, arising both from geographical proximity and from Internet-based organisational proximity. Particular relevance is given to policies aimed at facilitating the increasingly important role of "created assets", such as:

(a) Entrepreneurship based on "global thinking-local acting" visions, able to identify opportunities and threats in the global business environment, and to consequently adapt, according to its particular strengths and weakness.
(b) Knowledge mobility and Knowledge Leadership.
(c) Dynamic capabilities of the organisations.

Chapter 2

A Strategy to Sustain the Development Processes of a "Smart Community" in the Digital Economy

In this chapter the author explores, at a conceptual level, the new regional foundations of economic performance in the Knowledge-based Network Economy, not only for the region as a whole, but also for the single firms located in that region. More in particular, the author focuses on how the KBNE shapes the relationship between the innovation processes and the regional level of economic organisation. From this point of view, two main factors shape the new role of regions in affecting the innovation processes:

- The mobility, across national boundaries, of firms' created assets due to the globalisation processes which the world economy is undergoing.
- The increasing rate of learning processes inside organisations due to the emergence of a Knowledge-based Economy.

In the evolutionary approach followed by the author, both globalisation and KBNE are the result of the structural changes that the world economy is undergoing due to the new ICT techno-economic paradigm.

The conceptual framework emerging from this exploration has been used to set up a model of "Smart Community" to sustain regional innovation processes in the KBNE environment. A Smart Community is a geographical area ranging in size from a neighbourhood to a multi-county region whose citizens, organisations and government institutions "use" ICT to strategically position that region for economic, social and technological development.

The author introduces the Smart Community model as a consequence of the region-enterprise metaphor. This metaphor is based on the assumption that with regard to innovation performance and to knowledge generation, the dynamics governing an enterprise and a regional innovation network are essentially the same. Thus, in setting a strategy to sustain the creation and maintenance of innovation networks, one should take into account the new forms of enterprise organisation, i.e. the Internetworked Enterprise, and the enabling role of digital innovation in fostering networking.

Two main elements characterise a Smart Community:

- *Connectivity.* The extent to which communities can take advantage of ICT is directly affected by its degree of "connectivity" and thus of "virtualness". Connectivity refers to the access to the global digital network that includes satellites, cable TV, new wireless systems, mobile telephones and the Internet. Here, the adoption of ICT, and particularly of a large-bandwidth digital network, has two main objectives: (a) to support the learning and knowledge creation processes at intra- and inter-organisational level; (b) to facilitate the coordination mechanisms among different actors in a region. Digital networks allow implicit coordination by sharing information resources, access to common data repositories, and the use of Decision Support Systems (DSS) operating in a network configuration.
- *Increasing returns.* The evolutionary approach followed in this analysis highlights some form of increasing returns generated by network externality effects. Indeed, learning and knowledge creation are interactive and collective processes that involve the creation of "regional innovation networks" in which the functional boundaries within firms are porous, as are the boundaries between firms, and between firms and local institutions such as trade associations, universities, etc. In these networks, firms compete but also learn from each other through informal communication and collaborative processes.

Chapter 3

A Knowledge Hub for Regional Development

In the third chapter, "A Knowledge Hub for Regional Development", Mario Marinazzo analyses the growing complexity of the regional innovation landscape. He examines the conceptual basis of business innovation and Knowledge Leadership, and focuses his attention on regional government authorities and on regional entrepreneurial communities as "forces which shape change". This enables him to translate the concepts and methods commonly used in firms into local government practice. He then recalls several contributions which delineate the key role played by knowledge as a primary factor in innovation and competitiveness, and formulates a schema of the cycle of leadership in regional innovation.

Within this logical framework Marinazzo depicts the main characteristics of a Knowledge Hub for regional innovation, a knowledge management tool enabling local communities to become more participative in their innovation strategies, and more efficient in managing global business processes. He gives a practical and operational perspective to managing knowledge at the regional community level, with a view to supporting innovation-related learning processes in local business and institutions.

Virtualness in the Knowledge Hub for Regional Innovation increases returns in:

- Analysing the experience of other regional communities.
- Generating valuable content by presenting comments and viewpoints from forums and e-mails.
- Involving more and more backward and forward linkages rooted in business and institutions populating the Knowledge Hub.
- Promoting cooperation among participants who desire to solve the same problem or seize the same opportunity.

Capturing increasing returns in virtual communities activated through the Knowledge Hub for Regional Innovation means:

- Drawing investors to the region, and expanding supply-chain networks including local business.
- Drawing more and more participants to dialogue about regional development, and generate participant-based content.
- Promoting clustering and cooperation among local private and public organisations, and their project-oriented relationships with external organisations.
- Gathering information and knowledge from participants, and building regional innovation-oriented projects and programmes.

Chapter 4

Net Economy Business Models for SMEs Operating in Fragmented Markets

In Chapter 4 "Net Economy Business Models for SMEs Operating in Fragmented Markets" Tommaso Massari explores new e-Business strategies and models for SMEs operating in highly fragmented markets. The readers are provided with a general overview of how the Net Economy impacts on strategies, business and technological models of industrial clusters operating in traditional sectors. The chapter emphasises the role of "federated distributed information systems" and "peer-to-peer" business models and technologies in forging and shaping the new business arena for SMEs operating in these sectors.

The first part of the chapter describes new trends in e-Business model design, based on a more distributed architecture of data, services and systems. Based on some preliminary assumptions, the work attempts to analyse potential e-Business scenarios for SMEs operating in traditional sectors. This part of the chapter describes an e-Business model — "OpenMarkets[1]" — developed by the eBusiness Management School in conjunction with other partners in response to an EC initiative and based on peer-to-peer architectures, to accelerate the impact of Net Economy-led transformations on European SMEs.

The second part of the chapter describes a distributed system of peer-to-peer web services applied to the Tourism sector — regarded as "information intensive" and highly fragmented.

The chapter closes with some observations on new competitive strategies for SME clusters operating in fragmented markets such as tourism and suggests a way to speed up SMEs' transition towards the Net Economy.

Chapter 5

Towards Digital Innovation Leadership

In Chapter 5, Giuseppina Passiante and Valerio Elia describe a new model of leadership for guiding innovation in an organisation competing in the digital economy, including both firms and regions. This leadership paradigm is considered as the capability to enhance the intellectual capital of the organisation, through a continuous and dynamic development, enhancement, and use of the three components of intellectual capital: Human, Structural, and Social Capital. But this aim is also achieved through an organisation-wide integration of the three components. Based on this dynamic framework in which intellectual capital plays a fundamental role, the authors suggest a rethink of the knowledge management concept in terms of Knowledge Leadership, conceived as the capacity to create a suitable context, enabling organisational knowledge to be created.

The authors then stress the dilemma of how is it possible to create the leaders that are able to manage this process, and then to foster the dynamic capabilities of a firm or of a region. First, some specific capacities of leaders are pointed out:

- Competencies for measuring knowledge management (people competencies, IT systems and databases, learning capabilities, management capabilities, intellectual property, management philosophy).
- A capacity to manage Human Capital, generating a wide variety of results: experiences, know-how, skills, creativity.
- A capacity to scan across different disciplines and industries as the key to see emerging conditions, paradigm shifts and opportunities from the competitive landscape of the Net Economy.
- A capacity to look at the whole system, not just at their parts, since the whole is a constantly changing field of connections, relationships and patterns of interactions.
- A capacity to create and use a strategic intelligence network for both qualitative and quantitative information.

These capacities are considered the basic concepts and principles of an experimental incubator of Knowledge Leadership, able to set up organisations skilled in creating, acquiring and transferring new knowledge and insights.

In this chapter, the authors describe the model of Knowledge Leadership Incubator that characterises this experiment, the "e-Business Management School" (eBMS), that is currently being carried out at the ISUFI-University of Lecce (Italy). The eBMS experiment is described in terms of:

- The eBMS strategy, that is its major goals and actions in time and space.
- The eBMS organisation, including:
    - the eBMS organisational structure and processes;
    - the eBMS resources, in terms of distinctive competencies, and physical/informational resources.
- The eBMS environment, as the network of relationships that it has developed.
- The eBMS performance, in terms of value generated by the experiment.

Some final conclusions are also offered where the follow-up to the experiment is announced.

1

Industrial Clusters in the Net-Economy: Empirical Evidence and Some Theoretical Approaches

Giuseppina Passiante

1.1. Introduction

Recently, industrial clusters have been defined as "networks of production of strongly interdependent firms (including specialised suppliers), knowledge-producing agents (universities, research institutes, engineering companies), institutions (brokers, consultants), linked to each other in a value adding production chain" (OECD Focus Group, 1999). This definition sums up a lot of different theoretical approaches to industrial clusters, which have been represented as (Steiner, 1998):

- Concentrated forms of economic activities with strong connections to the knowledge infrastructure (knowledge clusters);
- Vertical production chains of rather narrowly defined branches, where subsequent stages of production form the core of clusters (for example textile clusters with the different manufacturing stages: tailoring, design and cutting);

- Sectorial concentrations at different levels of aggregation (for example automotive or electronics clusters);
- Collections of firms with a common basic technology (biotechnology clusters);
- Firms with common demands or needs (for example eco-clusters).

All the theoretical approaches emphasise the linkages and interdependences existing between actors in the network of production that realise products/services and create innovations. These linkages and interdependences concern:

- Dynamic interconnections existing between a cluster's industrial structure, its corporate organisations, its local institutions and culture (Saxenian, 1996). Industrial sectors include the social division of labour, the extent and nature of links between customer, suppliers and competitors in a particular sector or complex of related sectors. Corporate organisation refers to hierarchical or horizontal co-ordination, centralisation or decentralisation and specialisation of tasks within the firm. Local institutions and culture contribute to creating and sustaining regular patterns of social interaction in a region. In the Saxenian approach, the three dimensions are viewed as closely interconnected, and their feedback mechanisms contribute both to creating industrial clusters, and to increasing the innovation rate in the regional economies where they are located;
- Interactions between four elements in a territorial system (Porter, 1999):
  - The nature of local firm strategy, structure, and rivalry;
  - Factor conditions, or the basic endowments or conditions on which local firms seek to compete, including both tangible assets (such as physical infrastructure) and intangible assets (such as information, logical systems, university research institutes);
  - Demand conditions, or the nature of local demand;
  - The presence of related and supporting industries, including suppliers and successful competitors, that create business infrastructure and spur innovation and spin-off industries.

  Porter frames these elements as four facets of a diamond (Fig. 1.1), and considers the interrelationships among them as the key engine of the industrial cluster's competitive advantage: in this perspective, a cluster is considered a geographically proximate group of interconnected and associated institutions in a particular field, linked by commonalities and complementarities.

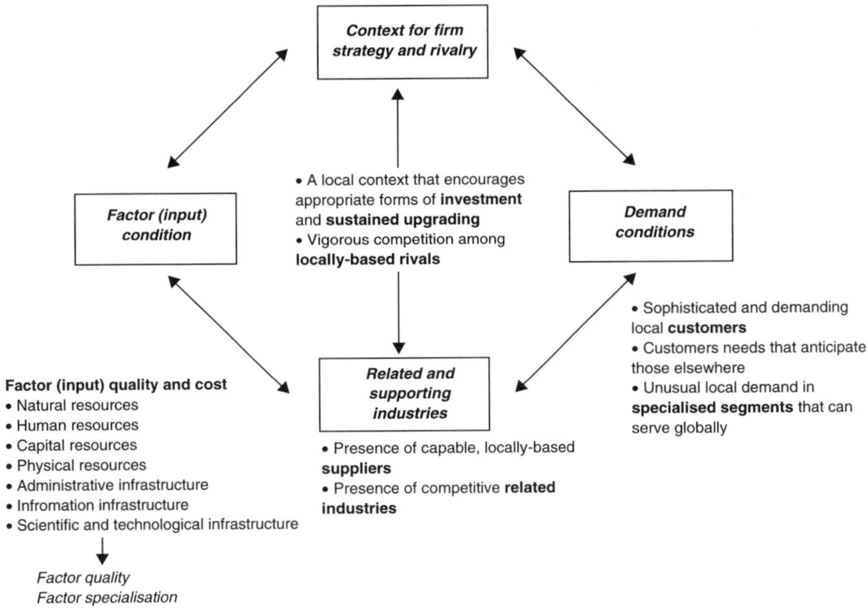

Fig. 1.1. Source of an industrial cluster's competitive advantage (Porter, 1998).

Clusters can be found in a variety of configurations, related to the specific context that characterises the dynamics of their interdependences. Examples of these contexts may be found in:

- The industrial districts of "the third Italy", specialised in traditional industries such as shoes, textiles, leather goods, furniture and ceramic tiles;
- Germany's Baden-Wùrttemberger mix of small and medium-sized makers of machine tools, textile equipment and automobiles components, alongside giant electronics corporation;
- The industrial cluster of the Rhône-Alpes region (France), specialised in advanced sectors such as electronics, ICTs and biotechnology;
- Flexible industrial clusters, identified in Denmark, Sweden and Spain;
- Canada's technology triangle (incorporating the four cities of Kitcheners, Waterloo, Cambridge and Guelph) where a cluster of technology-driven industrial activities is located, including telecommunications, computer software and hardware, environment technologies, advanced manufacturing, and the automotive sector;
- The Silicon Valley network-based industrial system which reflects localised social and productive interdependencies.

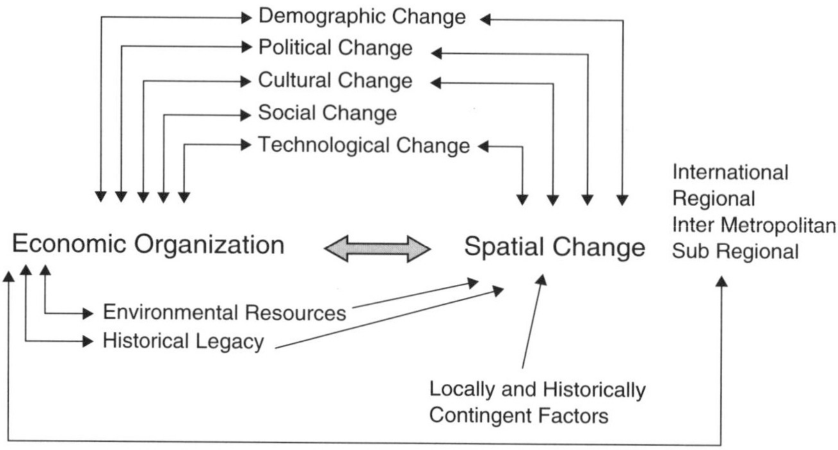

Fig. 1.2. The inter-relationships surrounding economic organisation and spatial change (Knox and Agnew, 1998).

In general, the nature of clusters and their evolutionary patterns could be related to the issues that historically characterise the interrelationships existing between the economic organisation and the spatial change of an economy. These issues are highlighted in the diagram suggested by Knox and Agnew (Fig. 1.2).

According to Knox and Agnew, the economic organisation of an industrial cluster approximates to the concept of production mode, i.e., the way in which human societies organise their production activities and thereby reproduce their socio-economic life. The "economic logic" of different modes of production determines different forms of spatial organisation of the industrial cluster. As feedback, the economic organisation of an industrial cluster is itself influenced by the local demographic, political, cultural, technological, and social environment, as both cause and effect (Knox and Agnew, 1998).

The industrial cluster's distinctive mode of production is a specialisation based on a sophisticated division of labour, that leads to interlinked activities and the need for cooperation. This "economic organisation" emphasises the role of spatial organisation based on geographic proximity. Cooperation needs closeness for just-in-time delivery, for communication, for the exchange of knowledge, especially in its tacit form. Cooperation between specialised actors in geographical proximity leads to spillovers and synergies (Steiner, 1998).

Geographical industrial clusters may be considered special patterns of economic activity related to organised or hierarchical capitalism. Radical changes in technology are crucial over time to produce new patterns of economic activities (Knox and Agnew, 1998).

The emerging electronically-networked world economy is creating new interrelationships between economic organisation and spatial change, related to the digital innovation paradigm. In this economic landscape a shift is emerging from the geographical industrial cluster to the virtual cluster, driven by digital innovation. The industrial clusters that are emerging in the Web-based world of business highlight a new interrelationship between economic and spatial organisation, a new competitive space where "How you do business" is more relevant than "Where you do business".

This chapter focuses on the analysis of the shift from geographic towards virtual industrial clusters. More specifically, the target issue concerns the changes that are taking place in the mechanisms of knowledge exchanges existing between the agents in a cluster, and the consequent impact on their innovative behaviours. Particular attention is devoted to the emerging global learning environment, driven by the "virtualness" concept.

The issues discussed in this chapter are organised into three sections:

- The first section is focused on an overview of the most seminal and consolidated approaches to industrial cluster analysis as systems of knowledge-based interactions, that increase their innovation rate;
- The second section offers some empirical evidence for describing and discussing the most important features of the emerging knowledge-based interactions in virtual clusters, and of their innovative effects;
- The third aims to point out the new global learning environment, which we have called the "Virtual Innovation System".

1.2. Some Seminal and Consolidated Approaches to the Analysis of Industrial Clusters

Several studies have focused on industrial clusters, and on their economic impact. The following non-exhaustive list includes some seminal approaches that are mainly focused on the knowledge exchanges between partners in industrial clusters. The aim is to gain some insights and thus compare the traditional role of geographic proximity in supporting the learning processes of a geographic industrial cluster with the analogous role of virtual proximity in a virtual cluster.

The list includes:

- Marshall-based approaches, that highlight the role of local social capital, as mainly pointed out by the "industrial districts" approach;
- Porter's approach, centred on the dynamics of geographic clusters in supporting innovation processes and competitiveness;
- The evolutionary approach, which emphasises the role of local knowledge spillovers and collective learning processes, mainly developed in the "milieu innovateur" approach.

1.2.1. *The Marshall-based approaches*

Marshall explains the development of concentrations of specialised industries in certain localities through the existence of positive externalities in agglomerations of interrelated firms and industries. The externalities are generated by three factors: knowledge spillovers between firms, specialised inputs and services from supporting industries, and a geographically pooled labour market for specialised skills.

To explain the specialisation and concentration of economic activities, Marshall goes behind exogenous comparative advantages or the international economies of scale. His hypothesis is based on the dynamic complementarities that develop within a system of interdependent economic entities. The relevance of knowledge spillovers is well expressed by his famous statement (Marshall, 1920):

> *When an industry has thus chosen a locality for itself, it is likely to stay there long: so great are the advantages which people following the same skilled trade get from near neighbourhood to one another. The mysteries of the trade become no mysteries; but are as it were in the air, and children learn many of them unconsciously. Good work is rightly appreciated, inventions and improvements in machinery, in processes and the general organisation of the business have their merits promptly discussed: if one man starts a new idea, it is taken up by others and combined with suggestions of their own; and thus it becomes the source of further new ideas. And presently subsidiary trades grow up in the neighbourhood, supplying it with implements and materials, organising its traffic, and in many ways conducing to the economy of its material.*

The knowledge spillovers existing within a system of interdependent economic entities both influence specialisation patterns in production and contribute to the growth of the overall system, generating knowledge spillovers for innovation and growth in other parts of the system.

The concept of dynamic complementarity, driven by the three factors that enable positive externalities, highlights how in Marshall the view of the innovation process is present in a cognitive space. Marshall's concept of knowledge spillovers between firms anticipates what has been suggested by more recent analyses: that the learning organisation does not store its knowledge in separate heads but instead stores knowledge in the relationships existing between stakeholders.

In the 1980s, several studies rediscovered and updated Marshall's work, highlighting the role of local interdependent economic entities in generating innovation processes. Here I refer mainly to the literature on industrial districts.

The industrial district approach mainly focuses on the relationship existing between industrial and regional dynamics. The basic features of the industrial district model may be summarised in the following stylised facts:

- Each industrial district originates from a specialisation either in a product or in a core ability (clothing, footwear, microelectronics etc). Production is divided into different phases, often compatible with small production units. These productive systems enable firms to reap all the benefits and savings from spatial agglomeration;
- Products and techniques are rooted in traditional know-how acquired and developed in the region;
- Close interdependencies between these local firms facilitate the flows of knowledge between them, thus leading to the development of production and innovation networks.

Indeed, industrial districts literature points out that these network organisations do not merely represent economies of scale, but are learning systems that develop via untreated interdependencies. Networks are generally based on strategic agreements, where firms know the results expected from co-operation: they are the results of a dynamic interdependence between the productive features of the firms and the social features of the population, mediated by the local social culture and prevailing institutions.

In this view, the region is considered a network of co-operative partnerships that continues the Marshallian lesson of positive externalities, and is identified as a set of localised links of knowledge, geographically concentrated and relatively self-contained, thus the region is no longer considered the scenario where the action takes place; rather, it is considered the action itself that generates innovations (Storper, 1997).

Already in this view, spatial organisation is related to the linkages and interactions between partners.

1.2.2. *Porter's approach*

Porter's conceptual framework places cluster issues in a broader and more dynamic view of competition, where location affects a cluster's competitive advantage by influencing its productivity, and especially its productivity growth (Porter, 1999). According to Porter, competition is dynamic and rests on innovation: "The productivity and the prosperity of a location rest not on the industry in which its firms compete, but 'how they compete"' (1999). In Porter's hypothesis, clusters affect innovation-based competition in three broad ways:

- By increasing the productivity of constituent firms or industries. Many of these advantages involve location-specific public goods or benefits that depend on face-to-face contact, close and ongoing relationships and "in-side" access to information, such as:
  - Access to specialised input and employers. As regards specialised inputs, a cluster member can get superior, or lower cost, access to components, machinery, business services, and personnel, as compared with vertical integration, or formal alliances, or by purchasing inputs from other locations. Moreover, sourcing within the cluster generally results in lower transaction costs, easier communications, lower cost of tailoring and facilitations in support services, such as installation, debugging, user training, troubleshooting and timely repair. As for experienced employees, Porter points out the advantages related to the lower search and transaction costs for recruiting, and to the more efficient matching of jobs to people, since the availability of specialised personnel usually far exceeds that provided by other locations and becomes a distinct benefit for cluster members;
  - Access to information. Porter correlates many of the competitive advantages of clusters to the free flow of information, the

  discovery of value-adding exchanges, the willingness to work across organisations, enabled by the relationships, networks and a sense of common interest that develop between the cluster members. In other words, supply and technological linkages and the repeated, personal relationships of a cluster, fostering trust, facilitate the information flow concerning extensive markets, technical information, and information about current buyer needs;
  - Complementarities between the activities of cluster participants, that cross products to create value for customers, not only in service delivery, but also in product design, logistics, after-sales services, and marketing. Porter stresses the concept of coordination and internal pressures for improving the members of a cluster, thus increasing its overall quality and efficiency, and enhancing the reputation of the cluster in a particular field, thus making it more likely that buyers will consider a cluster as a good vendor or manufacturer;
  - Incentives and performance measurement. As regards incentives, Porter highlights the role of competitive pressure, due to a constant comparison with local rivals who compete in similar general circumstances. Similarity is also considered a key factor for facilitating measurement of the performance of in-house activities;

- By directly fostering innovation, allowing firms to grasp opportunities related to:

  - A buyer's knowledge and relationships, and a faster plug into customer needs and trends;
  - Quicker perception of new technological, operating and delivery possibilities;
  - Higher flexibility and capacity to respond rapidly to changes in customer needs and to other opportunities for innovation, because of the juxtaposition of firms in related industries and the concentration of specialised information-generating entities;
  - Easier recruitment of specialised personnel to fill gaps, required to pursue new approaches;
  - Easier creation of complementarities in innovating opportunities at lower costs and the possibility of delaying large commitments until they are assured that an innovation will pan out;

- By supporting new business formation, related to:

  - Lower entry barriers, because assets, skills, input and staff can be assembled more easily by a new enterprise;

- Lower risk premium on capital requested by local financial institutions and investors, that already possess familiarity with the cluster;
- Multiple potential local market, often highly sophisticated.

In Porter's view, each of these broad influences of clusters on competition depends to some extent on personal relationships, face-to-face communications and interactions among networks of individuals and institutions. Indeed, many of the competitive advantages of clusters depend on the free flow of information. Porter's approach emphasises the central importance of a cluster's social structure. Economic activities are seen as being embedded in ongoing social relationships. The social relationships that constitute the social capital greatly facilitate access to important resources and information. Porter's recent approach bridges cluster theory to network theory and competition (1999). A cluster is conceived as a form of network that occurs within a geographical location, in which the proximity of firms and institutions ensures certain forms of communalities and increases the frequency and impact of interactions. The cluster offers a new way of exploring the mechanisms by which networks, social capital and civic engagement affect competition and market outcomes.

For many of these mechanisms, geographical proximity plays a strategic role. The spectrum of the mechanisms through which proximity increases cluster benefits is synthesised in linkages, transactions and information flows, which take place inside a time and a space (1999). Indeed, geographical proximity is a critical variable in Porter's theory of competitiveness, since it is considered a determinant factor to:

- Enhance the influence of the single determinants of the diamond, and their mutual reinforcement;
- Foster efficiency and specialisation, because of the concentration of competitors;
- Influence improvements and innovations in products and processes;
- Facilitate the entry of suppliers, users and related industries;
- Speed up information flows and innovation rates;
- Induce visibility of competitor behaviour;
- Lower transport and communications costs.

Therefore, in Porter's framework, clusters represent "spatial organisational forms" with close linkages with suppliers and buyers, which enhance the value of outsourcing or partnering, and determine demonstration effects, contagion, experimentation and ease of observability.

The relevance of geographical proximity in Porter's model may be observed in the following statements:

- Living and working in a circumscribed geographic area offers opportunities for repeated interactions, informal contracts, and advantages in transaction costs, fosters trust and open communication, and lowers the costs of severing and recombining market relationships;
- Geographically proximate clusters of independent and informally linked firms and institutions represent a robust organisational form in the continuum between markets and hierarchies.

1.2.3. *The evolutionary approaches*

Porter's approach highlights the role of clusters as a source of productivity growth and also of competitive innovation-related advantages.

The evolutionary theory points out that innovation, as the engine of productivity growth, springs from information asymmetries and market imperfections. Knowledge accumulation and learning mechanisms are thus considered as the key referring points of these approaches. The evolutionary theories suggest that a neural-net model of innovation is probably the most useful approach for interpreting knowledge accumulation and learning processes (Ziman, 1991). The patterns of nodes and connections in these cognitive spaces are not similar in different organisations or local systems. In each different organisation and local system, the learning engine lies in the particular configuration or pattern of ideas, techniques and commodities, and the specialised items of knowledge linking them. The learning capability of these transversal networks becomes their capacity to transform (De la Mothe and Paquet, 1998). The relationships between market and non-market organisations and institutions generate a context that Nelson and Winter define as the "selection environment", that is at the core of the processes of innovation, learning and discovery, and of the process of diffusion of technical and organisational innovations. In evolutionary approaches, the selection environment forms what is called "the relevant milieu" (internal or external, broader or narrower) that explains industrial clusters as innovation networks or as local and regional systems of innovation.

The notion of "milieu" connotes three sets of forces (Maillet, 1992):

- The contours of a particular spatial set, vested with a certain unity and tonus;

- The organisational logic of a network of interdependent actors, engaged in co-operative innovative activities;
- The organisational learning, based on the dialectics that develop between the internal milieu and the external environment.

At the core of these systems of innovation there are three profoundly interconnected processes: an adaptive and cumulative learning process, a system of interactions between proximate agents and groups and a dynamic coordination process built on the other two (De la Mothe and Paquet, 1998).

The evolutionary approach emphasises a dynamics of industrial clusters that has shifted from the economy as an allocative mechanism to the economy as a learning process (Lundvall and Johnson, 1994). The notion of learning economy and of innovation as a form of learning also allows us to rethink the perspective of industrial clusters as distributed networks of knowledge. The process of collective learning existing in clusters induces a stylised view of innovation processes in a cognitive space, that considers the knowledge accumulation process as path-dependent, non-linear and shaped by the interplay of market and non market organisations and institutions (Nelson and Winter, 1982).

In the dynamics of clusters, the best organisation for knowledge production depends greatly on a balance between communality and diversity of knowledge, between coherence and mutual learning, between exploitation and exploration (Marengo, 1993).

The learning process remains a social cognitive process, requiring interactions that come from geographical closeness, that generates not only maximal probability of learning by learning (i.e., of developing new capabilities), and not only through a greater density of situated cognition-driven interactions (Kirat and Lung, 1995). In this sense, spatial interactions are much less relevant *per se* than the mix of situated culture and institutions that characterises the context and facilitates communications, cumulative information exchanges and community learning.

In this perspective, proximity is not a solely spatial concept. Agents and groups may be proximate territorially, organisationally, and institutionally (De la Mothe and Paquet, 1998). According to De la Mothe and Paquet, since everything happens in time and in space, there may be a propensity to ascribe more causal importance to geographical or historical linkages than they really have, for the simple reason that co-relations are more easily detectable in those dimensions. Indeed, complementarities of knowledge, competencies and capabilities, or organisational and technological complementary connect[illegible] the causal factors through which faster-learning dynami[illegible]es can be explained (Richardson, 1972; Haas, 1995).

This "organisational proximity" is the variab[illegible]sensitive to the dynamic interrelationship existing between [illegible]ganisation and the spatial change of an economy, as suggeste[illegible]nd Agnew in Fig. 1.2. The technological discontinuity gene[illegible]nternet has supported the emergence of a new organisational proximity, that characterises a new competitive space, where "How you do business" is more important than "Where you do business".

The emerging virtual industrial clusters represent a new form of economic organisation, related to an "Internet-driven" spatial change.

1.3. Emerging Virtual Industrial Clusters: Some Empirical Evidence

The opportunities provided by the new digital economy landscape are determining radical changes in the organisational structures of firms, even the creation of new virtual clusterisation firms, i.e., distinct systems of suppliers, distributors, service providers and clients that use "internetworking technologies" as a principal way of co-operating and competing. These "virtual clusterisation firms", which have also been defined as "e-business communities" or "b-web communities", are defined here as "virtual clusters".

In a virtual cluster (VC), each enterprise adds one or more distinct aspects of product/service value to the value of the network, by exchanging digital knowledge with other members. The changes in supply chain management are related to the opportunities for business and industry to capitalise on the convergence of commercial interchanges, consumer markets, distribution channels and production units. The main virtual supply chain behaviours may be identified as:

- Horizontal collaboration among independent team-based structures, which is replacing traditional vertical hierarchies (Zenger and Hestley, 1997): teams behave as suppliers and customers toward other teams that are both internal and external to the organisation (Bartlett and Goshal, 1995);
- Production and transaction organisation based on inter-modality and complementarities, instead of substitution, based on the assumption

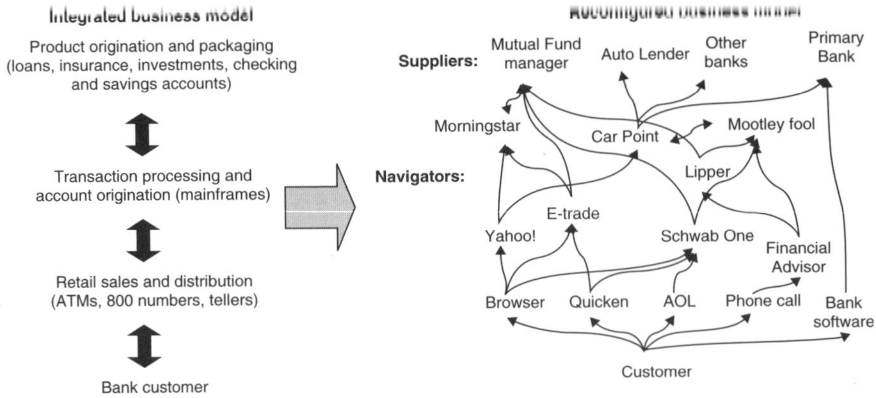

Fig. 1.3. The clusterisation process in retail financial services (Evans and Wurster, 2000).

that the best way to handle risk is to share it by leveraging the capabilities and resources of many players (Passiante and Andriani, 2000).

Recent empirical analyses focused on VCs highlight that the VC-enabling factors may be identified in ICT ubiquity (increasingly wireless) and bandwidth robustness, that allow firms to access real-time what they need and to co-ordinate their intra and inter-firm activities, creating value both by offering innovative and personalised products/services and by cutting transaction costs (Davis and Botkin, 1994; Rayport and Sviokla, 1995).

As an example, Fig. 1.3 shows the clusterisation process that has taken place in retail financial services. The old integrated business model, where multiple products were originated, packaged, and sold through a common set of distribution channels, with high fixed costs and substantial economies of scale, is evolving toward new clusterisation forms.

Indeed, today's reconfigured business model consists of small businesses that are largely stand-alone and compete product by product. Deprived of this traditional role, banks had to switch to the roles of navigator, agent and facilitator.

This new business model is the macroeconomic manifestation of a myriad of microeconomic processes, through which firms select their clients, differentiate their products, identify the activities they intend to realise within the company or that they outsource by developing new relationships with suppliers, distributors, clients and partners, to give value to their customers and make profits (Rayport and Sviokla, 1995).

Recent empirical studies highlight some common features of a VC:

- Internet Infrastructure (Tapscott, Ticoll and Lowy, 2000), to lower transaction costs: the internetworking technologies are changing the traditional Coase's trade-off. According to Coase's law: "a firm will tend to expand until the costs of organising an extra transaction within the firm becomes equal to the cost of carrying out the some transaction on the open market" (Coase, 1937). The Internetworking technologies, and their related virtual business models, have lowered the minimum threshold value that makes the outsourcing process advantageous, since they allow firms to reduce all the components of a transaction cost:
  - Search costs, both for finding information about the resources to be acquired and for determining when to trust a supplier. Indeed Internetworking technologies allow customers to use new "on-line" intermediaries, and more specifically directory services, both general and specialised in specific topics and search engines;
  - Contracting costs, because every exchange now requires a unique price negotiation and contract, without multiple trading and bargaining;
  - Coordination costs, that is the costs of coordinating the partners involved in the process of purchasing resources and realising and distributing products/services, because of the new opportunities of integrating the firm's information system with those of suppliers and clients/distributors;
- Five classes of participants (Tapscott, Ticoll and Lowy, 2000):
  - Customers, who receive and contribute to the value of the VC;
  - Context Providers, who support the interface between customers and suppliers. A context provider leads the value realisation and rule-making activities of a VC;
  - Content providers, who design, make and deliver the VC product/services;
  - Commerce service providers , who supply transactions and financial management, security and privacy, information and knowledge management, logistics and delivery services;
  - Infrastructure providers, who make the necessary Internetworking infrastructure possible;
- Customer centrality (Reicheld and Schefter, 2000). In VCs the gap between producers and consumers is blurring. As highly customised products and services replace mass production, producers must create

specific products that are imbued with the knowledge requirements, and tastes of individual customers. In VCs, consumers become involved in the actual design process. This perspective of blending production and consumption is reminiscent of the notion of "prosumption" suggested by Alvin Toffler. Customers, not just suppliers, increasingly create value by drenching products with their knowledge; customers can disaggregate value and re-aggregate it from multiple suppliers, thus enhancing their power. Therefore emphasis is on customer relationships, as opposed to brand images. Companies collect information about customers that enables them to target their most valuable prospects more effectively, tailor their offerings to individual needs, improve customer satisfaction and retention, and identify opportunities for new products or services. But even as more and more managers begin to build strategies based on capturing information about their customers, a major challenge is under way that may undermine their efforts;

- An e-lance behaviour on the part of VC actors (Malone and Laubacher, 1998). The VC competitive environment requires firms to re-think their boundaries. The successful VC organisation has transformed from classic, closely held bureaucracy to a center-less entity that can spread across the globe. The center-less organisation accomplishes its mission using a rich set of evolving relationships with its stakeholders, both internally (employees, managers, board of directors) and externally (partners, suppliers, customers) under one common identity. Tasks are not assigned and controlled through a stable chain of management but rather are carried out autonomously by independent contractors. These electronically connected freelancing, e-lancers, join together into a fluid and temporary network to produce and sell goods and services. When the job is done, the network dissolves, and its members become independent agents again, circulating in the economy, seeking the next assignment. All these trends point to the devolution of large, permanent corporations into flexible, temporary networks;
- Rules and standard (Tapscott, Ticoll and Lowy, 2000): VC participants know and adhere to the rules of engagement of their community;
- Knowledge (McDermott, 1999; Storck and Hill, 2000): VC partners use Internetworking technologies to exchange a variety of data, information and knowledge;
- Innovation(Sawhney and Prandelli, 2000; Tapscott, Ticoll and Lowy, 2000): a tool for generating value to the customer, related to an

integration of the values generated by each VC partner. The risk of a VC partner behaving opportunistically is offset by the advantages coming from integration, in terms of costs, speed and quality.

These features configure VCs as new systems of knowledge-based interactions, i.e., new cognitive spaces that support innovations within and between their individual participants.

In this paragraph, I aim to point out some issues concerning the innovation processes of these VCs, in order to compare geographic and VC innovative behaviours. To this end, the following paragraphs are focused on:

- A description of a typical configuration of a VC partner;
- A discussion of the main results of empirical research into the learning processes that the VC partners develop with their customers and through their strategic alliances, here considered the main source of their innovation processes;
- A presentation of the emerging VC learning environment, considered to be their new innovation system.

1.3.1. *The typical configuration of a VC member: the Internetworked Enterprise (IE)*

result of an integration process of different "core competencies" owned by each single partner, supported by the necessity to face the risks, costs and complexity of innovation. The final structure of a VC is then similar to a "hub and spoke" configuration (an example is shown in Fig. 1.4), that consists of many different nodes, interconnected by a web of linkages (transactions/relationships) (Passiante and Andriani, 2000).

Each VC node takes the shape of an "Internetworked Enterprise (IE)", internally connected via Intranet, with suppliers and customers via business-to-business networks and with other organisations, business homes and consumers via public Internet. Links include (Passiante and Andriani, 2000):

- Upstream transactions with suppliers, using Extranet solutions for managing and coordinating suppliers, or managing the supply chain;
- Downstream transactions with distributors and clients, allowing users to access information about credits, sell reports, products/services or to monitor transactions and orders or to get financial information;

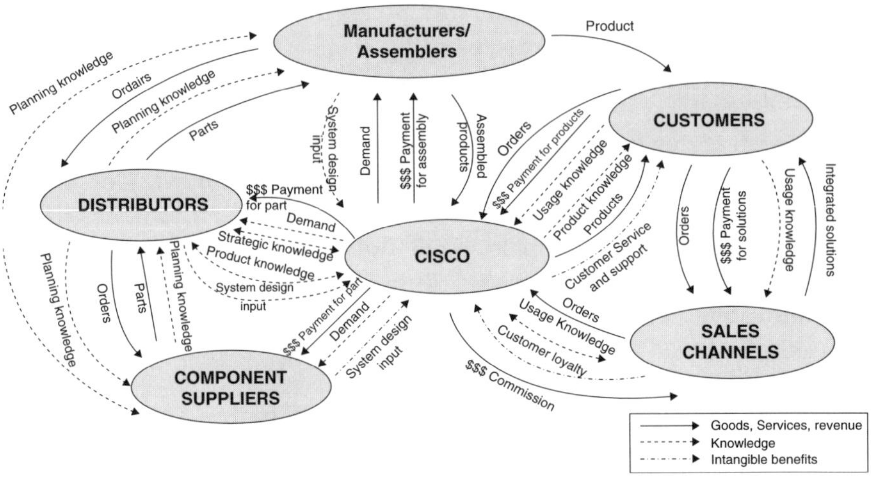

Fig. 1.4. A topology of a virtual cluster (Tapscott, Ticoll and Lowy, 2000).

- Horizontal transactions with competitors or other institutions, where the target is co-ordinating hardware/software manufacturers, venture capitalist and marketing information systems.

This configuration allows the IE to behave as a "sense and respond" organisation. Indeed, unpredictable, discontinuous change is an unavoidable consequence of doing businesses in VCs. This environment demands fast and sometimes instantaneous response, and as a result many companies are fragmenting themselves into smaller units in order to respond in real time. In this ever-evolving environment, business leaders need to be able to change and adopt new business models instantaneously. Organisations embrace not only adaptive mindsets, as in geographic clusters, but also adaptive business platforms. These platforms allow firms to deliver a "Sense and Respond Enterprise", as defined by Evans and Wurster (1997), also known as an "event driven organisation".

With a simple, adaptive, holistic and contextual solution, IEs have the ability to offer value and service to all stakeholders. Collaboration is typically defined as "working jointly with others, especially in an intellectual endeavour". In order to truly collaborate, shared meaning, processes and commitment must exist to create an "intellectual endeavour". This concept of sharing leads to a new wisdom based on the collective capabilities of an IE. Today's market leaders demand tightly integrated relationships with their suppliers, customers and partners to

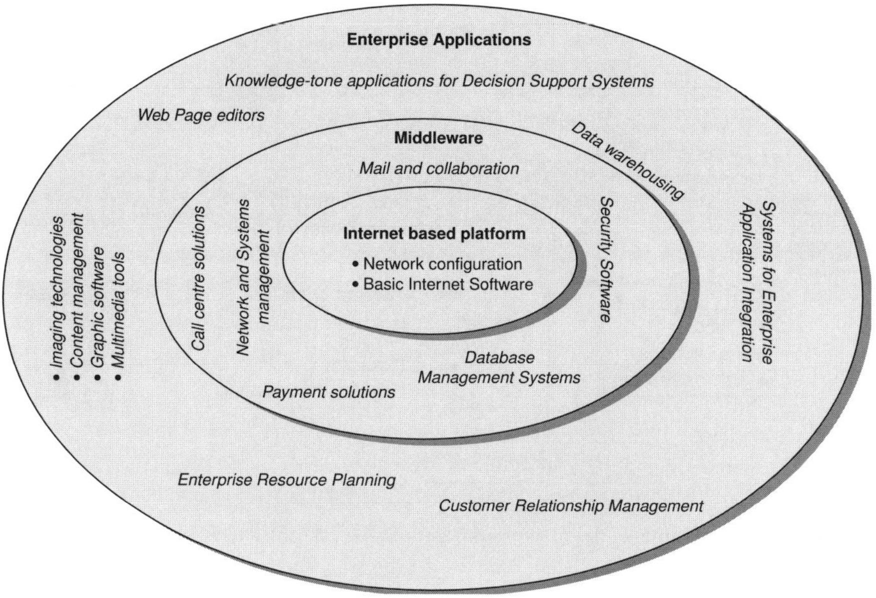

Fig. 1.5. The digital infrastructure of an Internetworked enterprise.

speed up innovation, build lasting relationships, increase productivity and raise profitability. The primary driver of growth and value is the IE's ability to collaborate across these relationships. More than ever, business partners need to share not only information, but also a more highly evolved capacity of working towards shared goals and efficiencies.

In order to collaborate with different partners and with different organisational assets, the IE exploits the opportunities coming from its digital infrastructure, that may be modelled (Fig. 1.5) in terms of:

a. The Internet-based platform, that includes:
   (i) Network configurations, such as the one shown in Fig. 1.6;
   (ii) Basic Internet Software, encompassing:
      i. Operating systems;
      ii. Software for implementing Internet protocols and for connecting to the services offered by the web-server;
b. Middleware applications, including:
   (i) E-mail systems, both private (that are used in companies and run by themselves) and public (used by individuals and mainly hosted by service providers);

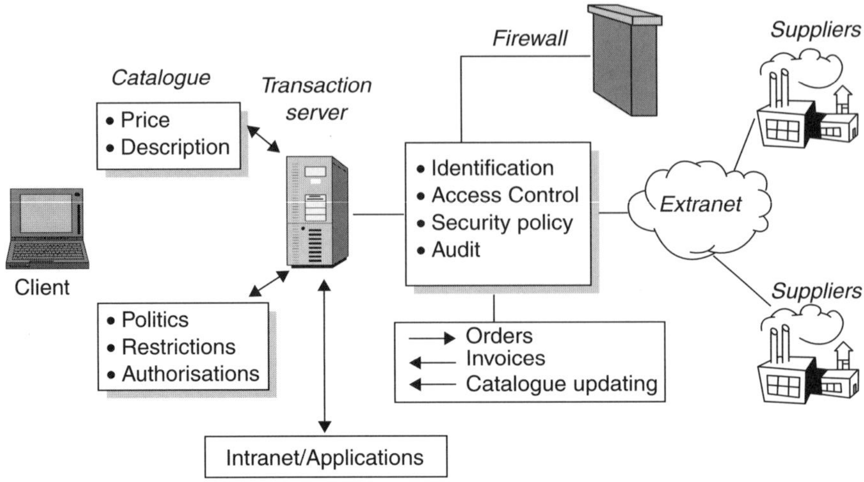

Fig. 1.6. An example of the network infrastructure of an Internetworked enterprise.

(ii) Network and system management applications, that allow the IE to manage all the networks, systems, desktops, databases and specific applications;
(iii) Security software, that supports the privacy, confidentiality and integrity of business transactions;
(iv) Payment solutions, to make transactions reliable, open to the different requirements of the customers, and flexible;
(v) Database management systems, to store large portions of data in a structured way;

c. IE-specific applications, such as:

(i) Imaging technologies;
(ii) Web-page editors;
(iii) Content management applications;
(iv) Graphics software;
(v) Multimedia tools;
(vi) Knowledge-tone applications, such as data warehousing;
(vii) Call centre solutions;
(viii) CRM (Customer Relationship Management) applications;

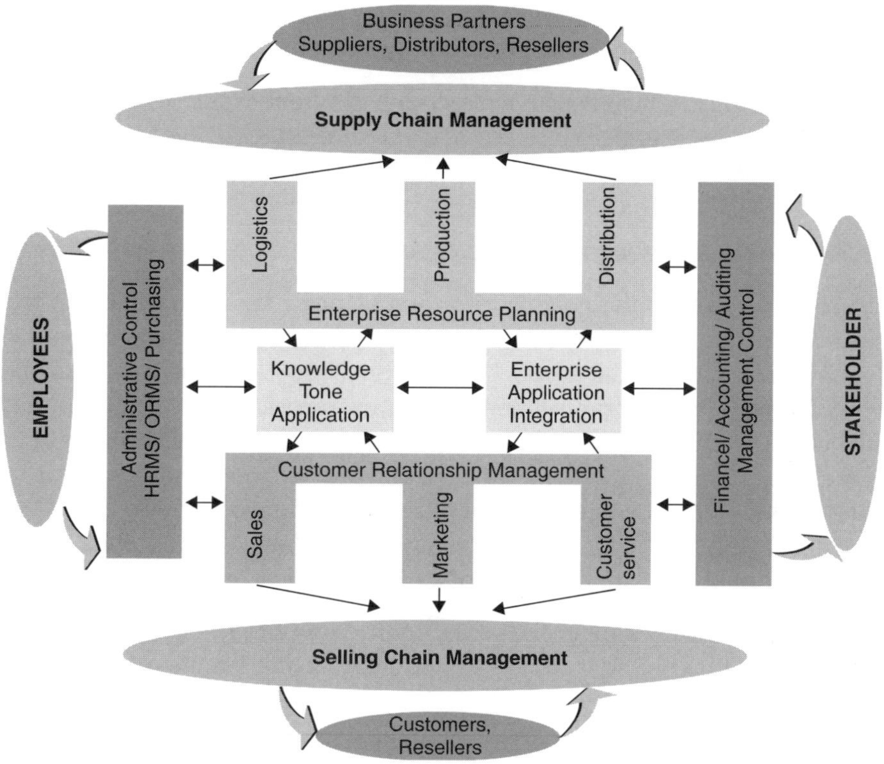

Fig. 1.7. The main relationships of an Internetworked enterprise (Kalakota, 2000).

(ix) ERP (Enterprise Resource Planning) applications;
(x) Systems for integrating the IE's applications.

These applications allow the IE to develop different relationships with its partners. Figure 1.7 gives a general framework of the main relationships that the IE develops by using its Internet-based Infrastructure.

These relationships are changing the internal assets of the IE, which is becoming a new type of entity, a less discrete enterprise that features a set of common activities in the midst of a network of relationships, grasping the opportunities coming from (Pekar, 2001):

- The technology advantages, that have blurred the distinct industries and facilitated open communication between partners;
- The possibility to develop cross capabilities, which the single firm is not able to develop, in order to satisfy new customers' needs;

- Additional degrees of strategy freedom;
- The possibility to speed up the learning curve.

This new entity, which is able to assemble different expertise rapidly, has enormously increased the timing and the amplitude of its learning rate (Quinn, 1999).

1.3.2. *The IE's learning processes*

In order to gain some insights into the characteristics and effects of the learning processes developed by IEs, I carried out an empirical research.

According to Nonaka and Nishiguchi (2001), analysis has been focused on the inter-organisational learning process, which is considered the key driving force for innovation processes, since it is increasingly rare for a single firm to possess all the financial and technological resources to undertake, alone, the development of a new product or service.

According to Tidd, Bessant and Pavitt (1997), the main inter-organisational learning processes of a firm have been grouped into:

- Market-related learning processes, connected with understanding user needs and involving lead users, who increase the likelihood of success of a new product/service;
- Learning processes built up through alliances with suppliers, competitors and other sources of knowledge. These networks of relationships are considered the most appropriate units of analysis for understanding a firm's innovation processes.

The analysis has included six case studies, identified through specialised information sources, integrated with Internet-based analyses. Cases were chosen in order to include the following (the classification is borrowed from Tapscott, Ticoll and Lowy, 2000):

- "Agorà" communities. "Agorà" is the electronic version of the primitive town market, aimed at providing partners with up-to-date information and news, as well as secure transactions. Many participants can bring goods to market or decide what the price should be. Price discovery mechanisms include one-to-one haggling, multiparty auctions and exchanges. "Fineco on line" was chosen as an "Agorà" case study;
- "Aggregation" communities, where a primary company organises and represents diverse product/service offerings to end consumers, positioning itself as an intermediary between producers and customers. Aggregators typically offer a diverse portfolio of products and

services, which often have little or no need to be fully integrated. "IBS" and "Wine" were chosen as "Aggregation" case studies;

- "Value chain" communities, where a primary company leads suppliers and distributors to achieve operational excellence in key distribution processes. The primary company's end is to produce a product/service outcome that pulls together the contribution of multiple resources in a single customer offering. This community is then built around the relationships between suppliers of raw materials and finished components, manufacturers and assemblers, distributors, retailers and customers, for cost and time savings. "Elettrodata" was chosen as a "Value chain" case study;
- "Alliance" communities, where multiple producers comply with standards, tailor and integrate their own solutions to create and support a "value space", i.e., an idea or vision of how to meet customers' requirements in a specific domain. "Borland" was chosen as an "Alliance" case study;
- "Distributive network" communities, aimed at facilitating the exchange and delivery of information, goods and services. Distributive networks typically provide commerce or infrastructure services to customers, context providers and content providers. "DHL" was chosen as a "Distributive network" case study.

The analysis started by using Internet browsers and search engines to identify the Web site of the main VC partner, and was then divided into two phases:

- Identification and analysis of the main partners' Extranet architecture, in terms of applications and links, characterised by using the information/documents available on the main partner Web sites;
- Identification and analysis of all other partners' Extranet architectures and links, singled out by scanning the Web sites of the other partners. Information was verified and integrated with primary data obtained via e-mail, participation in discussion lists and virtual communities.

The benefits of this method of research concern:

- The speed of the information search: by moving around ("navigating") the Internet it is possible to dramatically compress information acquisition time;
- The quality and the updating of information, because of the control of on-line content by VC partners;

- The easy accessibility of data without regard to geography, thanks to the universality of the World Wide Web standards.

Among the disadvantages, there are:

- The bias of information provided by VC partners, that cannot easily be verified;
- The fragmentation and overload of information available on the web, which needs careful selection and integration.

The following section presents the main results of our research.

1.3.2.1. *IE market-related learning processes*

The results of our empirical analysis highlight some frequent processes through which an IE enables learning processes with customers. Indeed, besides market surveys, our results show systematic processes through which information is obtained, different from information obtained from the traditional market surveys, which can be summarised as:

- Creating on-line communities, on-line forums, newsgroups, and discussion groups for monitoring the satisfaction level of customers, and for getting information about the opportunities to improve its products/services;
- To get quick feedback through emails from customers about any inefficiencies and dislikes in the services offered, and to inform customers when new products and delivery systems are introduced;
- To get information, through registration forms and by monitoring customer search patterns, in order to define the customer profile.

Differences in the IE learning processes spur from the strategic targets of the IE that is the leader of the cluster.

If the target is to improve the use of the existing products/services, and to meet known customer needs, as happens when products are in the maturity phase of their life-cycle, innovations mainly concern improvements in customer satisfaction levels, mainly in quality and delivery time. Products and services are differentiated on the basis of packaging, pricing and support, as happens in Wine, where the VC leaders interact with their customers in order to:

- Identify new customer requirements, and any major dislikes;
- Rank requirements according to their importance;
- Translate requirements into measurable characteristics;

- Establish the correspondences between customer requirements and the technical characteristics of the products;
- Choose appropriate criteria for measuring and defining value.

The analysis of customer requirements is then integrated with:

- Competitors' benchmarks;
- The creation of internal workgroups, involving lead customers, in order to:
  - Identify customer segments that are not adequately satisfied;
  - Imagine new solutions for increasing the efficiency and the effectiveness of the IE processes, through brainstorming and other techniques for human resource management, based on a wide use of Intranet-based tools;
  - Define and evaluate new possible solutions, by using techniques of quality management.

When the target is to serve new markets with novel products and services, as when products are in the "development phase" of their life-cycle, the key success factor of an IE is related to its capacity to evaluate/forecast the advantages that the new product is able to offer, compared with existing solutions, by identifying target customers and their purchasing behaviours. For example, in Elettrodata or in Fineco, learning-from-market processes are based on identifying the determinants which led target customers to choose a new product (lower costs, better performances, greater reliability). Often IEs develop strong co-operations with representative customers, as in Borland, which participates in the LINUX community in order to evaluate and forecast the diffusion rate of this new operating system, tanking information:

- About the peculiar characteristics of LINUX perceived by the customers: why they consider LINUX a better operating system in terms of its level of consistency with customers' culture and values, complexity, and visible facilities;
- For forecasting its future patterns of adoption, related to market and technological conditions.

To this end, Borland carries out customer and market surveys, on-line brainstorming, forums, chats, and expert opinion analysis, involving suppliers, dealers, customers, consultants and academics.

Generally, the learning process that emerges in these case studies has some common characteristics, such as:

- High rate of information knowledge exchange between the IE and its customers, because of the complexity of its products;
- Long-term co-operations, where the purchase happens a long time after the first information exchanges.

Learning processes between the IE and its customers also differ according to the range of different applications of the new product/service and the number of potential users:

- In the case of few applications and few users, direct face-to-face interactions are preferred to on-line ones;
- In the case of few applications, but many users, there is a careful segmentation, and interactions with users are developed using ICTs;
- In the case of many applications, but few users, IEs try to get information about conflicts and needs. The user-producer interaction is thus finalised to avoid an excess of customisation for one type of client;
- In the case of many applications and many different users, the learning process develops through multiple models of customers, in order to realise a product for the most generic market possible.

The learning process is thus time-consuming and complex, and is strongly influenced by:

- The rules and the standards of the VC;
- The degree of empowerment of the customer, and his personal role and responsibilities.

1.3.2.2. *IE learning processes through alliances*

The learning processes that firms develop through alliances are in general related to the opportunities for reducing costs, time and risk of development or market entry (Tidd, Bessant and Pavitt, 1997).

In the current competitive environment, characterised by high market uncertainty and high rates of technological change, the IE's propensity towards co-operation has evolved, reaching a stage when its value chain fragments into different independent businesses, each one owning its particular sources of competitive advantage, but grasping the opportunities of economies of scale at a VC level.

The results of our analysis show that even if the decision to co-operate is determined by the motivations and preferences of the individual partners, VCs develop differently according to the nature of the technologies and markets involved and, more specifically, to their complexity and confidential nature, with a greater propensity on the part of the IE to cooperate if:

- Technologies are complex, strategic for gaining competitive advantages, not easily transferable, and realised by highly reliable organisations, as happens in Borland, who have realised partnerships with Sun and IBM for sharing complex technological platforms;
- The IE has a corporate strategy favourable to technological leadership, internal capacities and technical know-how, is highly receptive to knowledge/information coming from outside, has a top-management that is familiar with new technology, has been successful in the past and has a high propensity towards risk. This is the case with Fineco, which has realised a wide spectrum of partnerships to get competencies both financial (a partnership with Bridge to obtain information about the trends of the financial markets) and technological (alliance with TIM for wireless technologies), because of the presence of a top management which is well disposed towards new ICTs.

The main factors which, by interacting dynamically one with another, determine the performances of an alliance may be grouped into:

- Motivation, both at a strategic level (such as Fineco's propensity for technological leadership) and at an operational level (as in Wine which has a strong propersity to reduce costs and timescales in their delivery processes);
- Technological factors, concerning the competitive significance, the complexity and the codifiability of the technology to be developed: these factors are evident in Borland;
- Organisational factors, related to the appropriateness of the IE's in-house competencies, its company culture, and its comfort with new technology, as the DHL case study highlights;
- Patterns of collaboration: here it is useful to point out the positive effects obtained in Fineco by using appropriate criteria for selecting partners, a good trust and communication level, clearly defined objectives and responsibilities;
- Interest in learning-by-interacting processes, high receptivity towards new knowledge and clarity of the IE partners' intentions, above all evident in the Borland case.

Moreover these factors influence the choice of the form of collaboration that is to be developed between the different options available (the taxonomy is borrowed from Tidd, Bessant and Pavitt (1997));

- Subcontracting (as happens in Wine and IBS), i.e., short term relationships concerning the outsourcing of non-core activities, that allow costs, risk and lead time to be cut, but generally reduce the performance and quality level of the final product;
- Technology licensing, as between IBS and Microsoft, i.e., fixed term relationships aimed at exploiting the intellectual property of other firms/organisations, in return for payment of a fee and royalties based on sales. Licensing allows quick access to the knowledge required for product development or for setting up new processes, but often reduces the IE's control of decision making;
- Strategic alliances, i.e., flexible agreements between two or more firms, to co-develop a new technology or a product (as between Fineco and TIM);
- Joint ventures, (as between Borland and Sun), i.e., long-term relationships to develop a new technology, a new product or to enter a new market. Joint ventures allow the know-how of individual partners to be integrated and enable managers to be employed full-time in the innovation process.

Typically, at the beginning, an IE develops strong relationships with a small number of primary suppliers, which share knowledge and competencies. Organisational linkages are related to strong personal on-line and off-line relationships between key individuals in the single IEs. In some cases, linkages evolve into a broader VC that includes secondary and tertiary suppliers and distributors, each contributing to the development of a subsystem of component technology.

VCs present two characteristics that affect the innovation process:

- The presence of rules and standards, which increases the efficiency of the VC, but creates barriers to change;
- Instability — the interdependencies existing inside a VC have an inherent instability and variability: this characteristic forms the basis for the learning opportunities that a VC creates for its partners. As a consequence, different VCs may represent different learning opportunities for an IE.

Moreover, the learning process that develops between the VC partners is boosted by a wide spectrum of factors, mainly by:

- The level of trust and communication among the partners (Handy, 1995);

- The importance of the VC as perceived by its partners (Evans and Wurster, 1997);
- A clearly defined vision and responsibilities agreed by all partners (Armstrong and Hagel, 1996);
- Alignment with the expectations of individual partners in the VC, (Malone and Lombarcher 1998);
- Equality in contributions and benefits among VC partners (Davis and Botkin, 1994);

Finally, the strategic position of the IE in the VC is also relevant. This position reflects its power and influence in terms of technological competencies, expertise, confidence, development rates, that allow it to access to a wide spectrum of information, through direct and indirect relationships, with different communication channels and different formalisation levels.

1.3.3. *Some stylised facts of the IE learning environment*

Learning processes, which each IE develops, determine new virtual clusterisation forms, that tend to develop around new sources of knowledge, instead of around geographical or spatial criteria, generating new collective learning processes, enabled by the Internetworked technologies. Indeed, the results of the empirical analysis corroborate some stylised facts about these new virtual clusterisation forms that recent studies have highlighted:

a) Internet acts as a general purpose technology, which allows enterprises to use any network technology, any type of middleware software and any type of application and service across the same network, to reach customers and suppliers worldwide (McKnight, 2001); Internet represents the strategic infrastructure of a new networked mode of economic organisation, as required by the need for cooperation and alliances, different from both trade (markets) and multinational firms (hierarchies) (Romano, Passiante and Elia, 2001). This networked mode of economic organisation exploits some technological capability dimensions of the Internet platform:

   (i) Interactivity: because of the real time and on-line nature of the Internet, relationships between organisations and customers are becoming more interactive, enhancing new paradigms of product design and customer service;

(ii) Connectivity: the radical increase in connectivity enabled by the Internet is giving rise to new communication and co-ordination mechanisms, both across organisations and within groups of individuals;
(iii) Scalability and extensibility, in terms of the ability to incrementally improve hardware and software, empowering Internet networks, services and applications, as the number of users, value of traffic, and variety of uses increases;
(iv) Interoperability, because of the Internet Protocol's features. Internet interoperability makes it easier, cheaper and faster to aggregate information in novel ways. From a strategic perspective, Internet becomes a weapon for disintermediating established firms and industries. Indeed as these "plug and play" enterprises have spread, value has shifted from entities that own the intelligence to those that orchestrate the flow and the combination of intelligence, i.e., intelligent hubs that coordinate the interactions between a network of partners, suppliers and customers.

b) Internet platforms dramatically influence both the location and the mobility of the enterprises. As regards location, modern high-speed networks have decoupled the intelligence of the VC (Sawhney and Parikh, 2001):

(i) Back-end intelligence (intelligence needed to process, store, and display information/knowledge for different user tasks) has become embedded into a shared infrastructure at the core of the networks;
(ii) Front-end intelligence (intelligence needed to interact with users) has fragmented into many different forms at the periphery of the network, where the users are located.

Between these levels, Internet-based networks have increased the mobilisation of intelligence: indeed intelligence has become fluid and modular. Small units of intelligence float like molecules, forming temporary groups whenever and wherever needed to solve problems.

The mobilisation of intelligence has profound organisational implications: connected by Internet-based networks, different companies combine their capabilities and resources into temporary and flexible alliances, to capitalise on particular market opportunities. More specifically, firms increasingly consider the creation of looser "cooperatives" and "constellations" as options to leverage different capabilities, some of which they are not even aware of or interested in exploiting at the time the relationships are established. In this way, we are moving toward a "centreless corporation" — an organisation in which competitive strength is based on harnessing capabilities and knowledge. This corporation selects, builds, and deploys the critical capabilities that enables it to gain competitive advantage, enhance customer value and drive its markets.

c) The emergence of a new Internetworked Enterprise, as a typical configuration of each node of a VC, suggests that VCs should be analysed in a link space, rather than in a geographical space (Romano, Passiante and Elia, 2001). More specifically, it suggests re-thinking the role of geographical space in generating opportunities for more effective and efficient access to information and knowledge, which are considered the key determinants in the innovation process. Traditional models relate these opportunities to a physical notion of distance and connectivity (Janelle and Hodge, 2001) that allows people, firms and institutions to access all the resources they need to innovate, prosper and compete.

   In VCs, the role of geographic proximity in supporting high rates of accessibility has become less strategic. The globally interconnected Internet is becoming an information resource that has lowered geographic barriers to building informed, knowledge-empowered societies, providing the potential to make all places equally accessible. Moreover, the Internet is reducing the importance of geographical location for determining interaction patterns, since activities have become more person-based, rather than place-based.

   Finally, more complex forms of organisation are emerging that, bypassing spatial relations, embed traditional places in broader networks of linkages, beyond traditional physical space. In this view, the geographic notion of space has been replaced by a new virtual space, that parallels the behavioural setting and rules of physical space with others that are based on electronic linkages between computers, allowed by shared hardware/software and by communications protocols. Activities in physical and virtual space are highly integrated: indeed virtual interactions have become both a substitute and a complement of physical interactions.

   Thus, a new, more generalised concept of proximity is emerging, related to a new concept of virtual distance, calculated as inversely proportional to the number of hyperlink connections between two points (web sites, organisations on so on). Geographic accessibility has thus been replaced by an information/knowledge accessibility, that can be measured by new parameters, such as (Dodge, 2001):

(i) Delays in response time among a set of dispersed computers, known as network latency;
(ii) Deliverability, related to the problem of data being lost in transit and having to be resent;
(iii) Availability of the network and servers, which is generally assessed in terms of the amount of downtime.

d) In the link space enabled by Internet platforms, collective learning processes are related to instantaneous shared communications, which trigger the knowledge conversion processes upon which learning processes are based. More particularly, interactions generate a continuous and dynamic tacit-explicit knowledge conversion among individuals, groups and IEs.

In order to explore the contribution of the Internet based platform to these interactions, every conversion process of tacit and explicit knowledge has been associated with some specific IE applications, as shown in Table 1.1 (Romano, Passiante and Elia, 2001; Passiante and Andriani, 2000).

IE applications such as videoconferencing (desktop video) and room based videoconferences enhance the learning mechanism related

Table 1.1. Correspondence between knowledge conversion processes and IE applications (Romano, Passiante and Elia, 2001).

	Tacit knowledge	Explicit knowledge
Tacit knowledge	*Socialisation* • Videoconferencing (desktop video) • Room based videoconference	*Externalisation* • e-mail • GroupWare shared databases • consulting • collaboration tools • workflow tools
Explicit knowledge	*Internalisation* • GroupWare (DSS) • GroupWare shared databases • data mining • search engine • web-casting	*Combination* • telephone • www based phone systems • e-mail, FTP directories, Usenet • audio-conferencing • chat systems • web chat • computer conferences • web-casting • document archive • workflow tools

to the conversion from tacit knowledge to tacit knowledge (socialisation process). In fact brainstorming camps, informal meetings, detailed discussions, sharing experiences between product developers and customers, facilitate the socialisation process. IE applications such as e-mail, GroupWare shared databases, consulting, collaboration tools, workflow tools enhance the learning mechanisms related to the articulation of tacit knowledge into explicit concepts (externalisation process).

IE applications such as telephones, www based phone systems, e-mail, FTP directories, Usenets, audio-conferencing, chat systems, web chats, computer conferences, webcasting, document archives, and workflow tools enhance the combination process, which represents the conversion from explicit knowledge to explicit knowledge: individuals exchange and combine knowledge through such media as documents, meetings, telephone conversations or computerised communication networks.

IE applications such as GroupWare (Decision Support Systems), GroupWare shared databases, data mining, search engines and webcasting enhance the internalisation process, which represents the embodying of explicit knowledge into tacit knowledge. This conversion is helped by knowledge verbalised or diagrammed into documents, manuals or oral stories.

Raising the level of analysis further, from the IE learning environment to a VC level, where a VC is a network of IEs, we get the profiles of new patterns of economic activities related to digital innovation.

1.4. VCs as New Patterns of Economic Activity: Towards the Virtual Innovation System

Digital innovations related to ICTs are modifying the interrelationships existing between economic organisation and spatial change, and are generating a new, knowledge-driven, competitive environment.

Dunning highlights the principal characteristics of the new economic organisation (1997):

- The increasing complexity and specialisation of economic activity;
- The growing interdependence of many markets on intermediate products;

- The acceleration towards a knowledge and innovation driven economy;
- The widening territorial boundaries of firms;
- The increasing significance of created assets (e.g. human skills and technological capacity) in the value-adding process;
- The evolution of new institutions and organisational forms;
- A re-evaluation of alternative systems for allocating resources.

These factors impact on the relative advantages of markets, firms and governments as mode of governance.

The new economic organisation is generating a change in the spatial division of labour, in the regional economic specialisation and in the exploitation of economies of scale, agglomeration economies and localisation economies. Dunning highlights the principal features of this spatial change:

- The increasing mobility of firm-specific resources and capabilities, especially knowledge-related assets, across national boundaries;
- The growing significance of cross-border transactions, which are either intra-firm, or between firms with ongoing co-operative agreements;
- The dramatic reduction in long-distance transportation and communication costs and of the physical and cultural barriers between countries;
- The growing importance of location-bound assets, notably an educated labour force and a sophisticated physical infrastructure, in influencing the setting of value-added activities by SMEs.

The diffusion of Internetworked technologies is generating new intelligent link-spaces, characterised by the ubiquitous provision and distribution of information in networks of agents. According to Carley, intelligent spaces will have four main characteristics (Carley, 2000):

- Ubiquitous access: The agent (human or artificial) will have the technology to access or provide information wherever and whenever it is useful, thus acting and remotely enabling other agents to act;
- Large scale: Vast quantities of information will be automatically collected, stored, and processed by vast numbers of agents;
- Distributed cognition and intelligence: Information, access to information, and processing and communication capabilities (i.e., intelligence) will be distributed across agents, time, space, physical devices, and communications media;
- Invisible computing: The interface between the digital world and the analog world will become seamless as computers are miniaturised, made reliable and robust, and are embedded into all physical devices.

In these intelligent spaces, "virtualness" becomes a new source of competitiveness, since:

- It offers a context where value creation opportunities may result from new combinations of information, physical products and services, innovative configurations of transactions, and the reconfiguration and integration of resources, capabilities, roles and relationships between suppliers, partners and customers;
- It broadens the notion of innovation since these spaces span firm and industry boundaries; it involves new exchange mechanisms and unique transaction methods (rather than merely new products, or production processes); and it fosters new forms of collaboration between firms;
- It provides an alternative to ownership or control of resources and capabilities. The alternative is to access such resources through partnering and resource-sharing agreements. Hence, value creation becomes more challenging, because rivals may have easy access to substitute resources as well;
- It decreases the direct costs of economic transactions — coordination costs and transaction risks — and reduces indirect costs, such as the costs of adverse selection, moral hazard and hold-up.

"Virtualness" enables the pervasiveness of innovation: innovation becomes ubiquitous in every industry, in every place and in every firm, and transcends local, regional and national borders. Virtualness contributes to generating dynamic external economies, through which complementarities of knowledge and competencies or organisational and technological connections may accelerate collective learning processes, enabling innovative capabilities. The concept of Virtual Innovation System (VIS) was introduced in a previous paper (Romano and Passiante, 1997) as a new unit of analysis for describing the global learning environment, which enables a shift from the economy as an allocative mechanism to the economy as a learning process, from geographical industrial clusters to virtual industrial clusters. VIS dynamics highlight, on a global scale, the well-known stylised facts to be found in the innovation literature:

- Innovation as a non-linear process, network shaped and dynamised in a fundamental way by a complex multi-dialogue which weaves the various partners together (De la Mothe and Paquet, 1998);
- Innovation first and foremost as new practical knowledge, generated by the capabilities/absorptive capacities of organisations (Langlois and Robertson, 1985);

- The strategic choice that strikes a balance between exploitation of existing resources and exploration of new possibilities and opportunities, in a rapidly-evolving, surprise-generating context (Mertins, Heisig and Vorbeck, 2001) ;
- Knowledge creation related to a balance between communality and diversity of knowledge, between coherence and mutual learning, between exploitation and exploration (Marengo, 1993);
- Learning and discovery related to the interactive mechanism with the context environment through which selection occurs. This innovation mechanism "both provides the source of differential fitness — firms whose R&D turns up more profitable processes of production or products that grow relative to their competitors — and also tends to bind them together as a community" (Dosi and Nelson, 1994);
- Proximity that breeds interaction and socio-economic learning related to the "selection environment", shaped by market and non-market components (Nelson and Winter, 1982). Proximity is not a one-dimensional or solely spatial concept. Agents and groups may be proximate territorially, organisationally, institutionally (Rallet and Torre, 1998).

The VIS may thus represent the global context where IEs and VCs develop the dynamics that characterise the "Ba" concept (Nonaka, Toyama and Konno, 2001). "Ba" is defined as a "shared context", where knowledge is created, shared and diffused. The key concept that underlines "Ba" is "interaction". Knowledge generation is considered a dynamic process that develops through interactions among individuals, or between individuals and their learning environment. Ba is therefore considered as the context shared by those who interact with each other and, via such interactions, creates and shares knowledge. "Ba has a complicated and ever changing nature. It is a boundary for interactions among individuals, yet its boundary is open.... Ba is an open place where participants with their own contexts can come and go and the shared context (that is Ba) can continuously evolve. By providing a shared context in motion, Ba sets binding conditions for the participants by limiting the way in which the participants view the world. Yet, it also provides participants with higher viewpoints than their own" (Nonaka, Toyama and Konno, 2001).

"Ba" is thus framed as a platform of knowledge that provides the energy, quality and place to perform the individual's/group's/organisation's learning processes.

Because of these particular characteristics "Ba" seems to be more suitable to understand the VC learning environment, characterised as a VIS. Indeed, Ba has some further characteristics that are similar to the VIS:

- It is not necessarily related to a physical space: it is a concept that unifies a physical space (a place), a virtual space (an e-mail service) and a mental space (shared ideas);
- It develops at different ontological levels: individuals form a Ba (a group) and, in turn, groups form the Ba of an organisation; the interactions that develop at different ontological levels amplify the knowledge-creation processes of organisations;
- It acts as an autonomous, self-sufficient unit, that may connect to other Ba's in order to expand its knowledge. The relationships among Ba partners are not predetermined: indeed, Ba is the result of spontaneous interactions that develop between its participants, rather than the effect of mechanistic planning;
- It has to be energised: knowledge participants need some specific conditions, such as:
    - Autonomy, which increases the commitment of individuals to creating new knowledge, and may be the source of unexpected knowledge;
    - Creative chaos, that encourages people to transcend existing boundaries to define and solve problems;
    - Intentional overlapping of information about business activities, to support the sharing of tacit knowledge and the definition of single members' roles;
    - Diversity, in order to deal with the challenges of the complexity and the variety of the environment;
    - Trust and commitment, since interactions among individuals participating in a Ba or between different Ba's have to be supported by trustful sharing of knowledge and continuous exchanges between all units of the Ba.

This suggests considering the VIS as a global context where IEs and VCs behave like complex adaptive systems (C.A.S.), with an internal dynamic of self-sustaining cycles of knowledge exchanges among their Internetworked organisations. In other words, a VC is characterised by positive self-sustaining knowledge mechanisms among its Internetworked organisations, which develop highly interconnected web of relations among them, generating the degree of "complexity" of the network (Passiante and Andriani, 2000). According to Passiante and Andriani, VCs highlight some characteristics of a CAS, such as:

- Coevolutionary systems. The term "coevolution" refers to "reciprocal evolutionary change" (Thompson, 1994), whereby the evolutionary trajectories of the species (organisations) in a system become interdependent over time. The action of an actor determines a series of cascading effects on the other actors that are interconnected. Connections can also feed back, creating a loop. In some situations, they may generate a destabilising positive feedback, that is, a mutual reinforcement of cause and effect, which, as in the chain reaction, can create the right environment for radical innovations (Stacey, 1995). Coevolution can be, to an extent, managed: in a VC this means managing the context of collaborative relationships in order to facilitate the emergence and operation of synergies (Eisenhardt and Galunic, 2000). This role in a "hub and spoke" structure is assumed by the central organisation;
- Non-linear systems, from two points of view. The first concerns the cause-effect relationship. There is no direct proportionality between input and output, due to the presence of a feedback mechanism in the system. The second concerns the value generated by the network, which cannot be deduced by summing the performances of the single actors, because it is strictly related to the synergies created by the network links. In other words, it is the dynamic interaction among actors, which exponentially increases the innovation opportunities provided by the entire network;
- Self-catalytic systems. A useful example of self-catalytic reaction is given by Kauffman (1995), concerning chemical reactions which can be self-catalytic in the presence of a sufficient variety of molecules: the auto-catalysis occurs when a product of reaction (say, X) catalyses its own production through a positive feedback process. The amount of X formed at any instant depends on how much X was there in the first place. In an analogous way, a VC presents a fundamental characteristic: the value of the network increases exponentially with the number of connections. In this way a VC may generate positive feedback (Shapiro and Varian, 1999) and a virtual cycle. The more popular the network, the more attractive it becomes for other users; on the other hand, it can lose value as it is abandoned by users;
- Path-dependent systems which have a history. There are no general laws that apply independently of the VC specific evolution: luck or

chance (or, in Prigogine's (1989) terms, "fluctuations") often play relevant roles. The single actor's diversity (Allen, 1997) is also a necessary condition for their evolution. Actors' interdependence generates their co-evolution, as an evolutionary process that Kauffman describes as a rubber landscape, where each step taken by an actor modifies the environment of all the others;

- Dissipative systems, which need a constant flow of energy with the external environment to survive. Like a CAS, a VC is an "emergent" system, and emergence requires order and a negative flow of entropy to support this order (Nicolis and Progogine, 1989). In VCs, energy is necessary to maintain up-to-date Web sites, to encourage knowledge exchanges among partners, to recruit new partners, to motivate stakeholders, to shake up the network, and to provide a new set of challenges that cannot be faced using existing procedures.

More specifically, VCs exhibit some of the characteristics of "edge of chaos" structures. Relevant properties include:

- Self-organisation on macroscale patterns based on repeated interactions on the microlevel;
- Continuous dynamic change of the internal structure by means of modification of locus and density of connections, either as an adaptation to external constraint changes or as a response to internal innovation;
- Emergent macroscale correlation among the nodes of the network;
- Emergent strategy.

1.5. Conclusions

The above analyses and discussions allow some concluding remarks to be made about the dynamics of the shift from geographical clusters towards virtual clusters, which may be synthesised in the following stylised facts:

- Internet-based digital connectivity is creating new interrelationships between economic organisation and spatial change. In this way, new patterns of economic activities are emerging. This new Internet-based "economic organisation" is determining different forms of spatial organisation among industrial clusters. The positive interference between organisational and geographical proximity is generating new link-spaces, characterised by the ubiquitous provision and distribution of information across networks of agents. "Virtualness" becomes a relevant source of competitiveness, because it enhances the advantages deriving from the multidimensional character of proximity, and introduces new ways and mechanisms of specialisation, of integration and of knowledge exchange;

- The basis of competitiveness comes from network organisations, mutually dependent relationships, globally oriented physical and communication infrastructure, knowledge creation and continuous improvement processes. "Virtualness" accelerates the pervasiveness of innovation; digital innovation, transcending local, regional and national boundaries, calls for new units of analysis, useful for representing the global learning environment in the new competitive landscape. Rather than ushering in the "end of geography", the VIS allows the globalisation processes to occur increasingly through regional systems' interdependence and integration;
- The VIS concept allows us to focus on processes rather than structures for analysing new patterns of economic activity and for comprehending the meaning of the shift from the economy as an allocative mechanism to the economy as a learning process. Marshall's concept of knowledge spillovers between firms maintains its value, because knowledge is stored in the relationships between agents in the VIS context. Moreover, positive externalities in agglomerations of interrelated firms and industries are not limited to certain localities, but assume a more dynamic form in intelligent link spaces;
- The intelligent link spaces that shape the VIS reduce the dependence of Porter's diamond from geographical constraints, and allow us to consider technological discontinuities as endogenous rather than exogenous factors of Porter's diamond. Moreover, the three broad ways that Porter considers the main influences of clusters on innovation-based competition, are enhanced in the VIS context, increasing the productivity of agents, enabling new opportunities to be grasped by innovation and supporting new business formation;
- VIS dynamics are characterised by learning mechanisms and knowledge accumulation processes that are the key issues in evolutionary approaches. Particularly, VIS dynamics are characterised by three profoundly interconnected factors:

  - An adaptive and cumulative learning process;
  - A system of interactions among organisationally proximate agents and groups;
  - A network logic.

These stylised facts lead to some policy implications, concerning the role of local/regional/national governments in fostering dynamic agglomeration economies, arising from geographical proximity and internet-based organisational proximity. Particular relevance is assumed by policies aimed at facilitating the increasingly important role of "created assets", such as:

a) Entrepreneurship based on "global thinking-local acting" visions, able to identify opportunities and threats in the global business environment, and consequently to adapt itself, according to its particular strengths and weaknesses;
b) Knowledge mobility;
c) Dynamic capabilities of organisations.

As regards the first point, some useful actions may concern support for:

- Global strategic alliances;
- Promotion of export through new media, such as Internet marketing;
- Specific efforts of local organisations to attract global service providers.

As concerns point b), the government could sustain:

- The gathering, management and communication of specific information/knowledge useful for local clusters;
- Local universities' efforts in customising technologies for local clusters;
- Pre-competitive research, by facilitating the access to foreign skills and technologies;
- A specialised communication infrastructure, to lower distance-related transaction and coordination costs, and knowledge asymmetries.

Finally, as concerns point c), some useful actions could be focused on:

- Basic ICT education for all, to narrow knowledge gaps;
- Generation of life-long learning opportunities for people;
- Development of technical skills, to allow people to grasp the opportunities offered by sophisticated digital technologies;
- Advanced education and training programs, aimed at supporting people in strategic management practices, suitable for competing on the edge in the digital economy scenario.

References

Allen P. (1997) *Cities and regions as self-organising systems*. Gordon and Breach Science Publisher, Amsterdam.

Armstrong A. and Hagel J. (1996) The real value of on-line communities. *Harvard Business Review* May–June.

Arthur W.B. (1996) Increasing returns and the new world of business. *Harvard Business Review* July–August.

Bartlett C. and Goshal S. (1995) *Managing Beyond Borders*. Harvard Business School, Boston.

Carley K.M. (2000) Organisational change and the digital economy: A computational organisation science perspective, in Brynjolfsson E. and Kahin B. (eds.) *Understanding the Digital Economy*. MIT Press, Cambridge, MA, 325–351.

Coase R.H. (1937) The nature of the firm. *Economica* **4**(4).

Davis S. and Botkin J. (1994) The coming of knowledge-based business. *Harvard Business Review* September–October.

de la Mothe J. and Paquet G. (1998) Local and regional systems of innovation as learning socio-economies, in de la Mothe J. and Paquet G. (eds.) *Local and Regional Systems of Innovation*. Kluwer Academic Publisher, Norvell, MA, USA, 1–26.

Dodge M. (2001) Accessibility to information within the Internet: How can it be measured and mapped?, in Janelle D.G. and Hodge D.C. (eds.) *Information, Place, and Cyberspace — Issues in Accessibility*. Springer Press, New York.

Dosi G. and Nelson R.R. (1994) An introduction to evolutionary theories in economics. *Journal of evolutionary economics* **4**(3).

Dunning J.H. (1997) A business analytic approach to governments and globalisation, in Dunning J.H. (ed.) *Governments, Globalisation and International Business*. Oxford University Press, London, 114–131.

Eisenhard K.M. and Galunic D.C. (2000) Coevolving at last: A way to make synergies work. *Harvard Business Review* January–February.

Evans P. and Wurster T.S. (2000) *Blown to Bits*. Harvard Business School Press, Boston.

Evans P. and Wurster T.S. (1997) Strategy and the new economics of information. *Harvard Business Review* September–October.

Freeman C. and Perez C. (1988) Structural crises of adjustment: Business cycles and investment behaviour, in Dosi G. *et al.* (eds.) *Technical Change and Economic Theory*. Pinter, London, 38–66.

Haas S. (1995) Economies externes technologiques, apprentissage et rendements d'agglomération, in Lazaric N. and Monnier J.M. (eds.), 206–227.

Handy C. (1995) Trust and the virtual organisation. *Harvard Business Review* May–June.

Janelle D.G. and Hodge D.C. (2001) Information, place, cyberspace and accessibility, in Janelle D.G. and Hodge D.C (eds.) *Information, Place, and Cyberspace — Issues in Accessibility*. Springer Press, New York.

Kauffman S. (1995) *At home in the Universe*. Oxford University Press, Oxford.

Kirat T. and Lung Y. (1995) Innovations et proximités: le territoire, lieu de déploiement des processus d'apprentissage, in Lazaric N. and Monnier J.M. (eds.), 206–227.

Knox P. and Agnew J. (1998) *The Geography of the World Economy*. Arnold, New York.

Langlois R.N. and Robertson P.L.A. (1995) *Firms, Markets and Economic Change*. Routledge, London.

Lesser E. and Everest K. (2001) Using communities of practice to manage intellectual capital, *Ivey Business Journal* **65**(4).

Lundvall B. and Johnson B. (1994) The learning economy, *Journal of Industry Studies* **1**, 23–42.

Maillet D. (1992) Milieux et dynamique territoriale de l'innovation. *Canadian Journal of Regional Science* **15**(2), 199–218.

Malone T.M. and Laubacher R.J. (1998) The Dawn of the E-Lance Economy. *Harvard Business Review*, September–October.

Marengo L. (1993) Knowledge distribution and coordination in organisations: On some social aspects of the exploitation vs exploration trade-off. *Revue internationale de systémique* **7**(5).

Marshall A. (1920) *Principles of Economics*. Macmillan, London.

McDermott R. (1999) Why information technology inspired but cannot deliver knowledge management. *California Management Review* **41**(4), 103–117.

McKnight L.W. (2001) Internet business models: Creative destruction as usual, in McKnight L.W., Vaaler P.M. and Katz R.L. (eds.) *Creative Destruction*. MIT Press, Cambridge MA, 39–59.

Mertins K. Heisig P. and Vorbeck J. (2001) *Knowledge Management — Best Practices in Europe*. Springer, Berlin.

Nicolis G. and Prigogine, I. (1989) *Exploring Complexity: An Introduction*. Freeman, New York.

Nishiguchi T. (2001) Coevolution of interorganisational relations, in Nonaka I. and Nishiguchi T. (eds.) *Knowledge Emergence*. Oxford University Press, London.

Nonaka I. and Nishiguchi T. (2001) Social technical and evolutionary dimensions of knowledge creation, in Nonaka I. and Nishiguchi T. (eds.) *Knowledge Emergence*. Oxford University Press, London.

Nonaka I., Toyama R. and Konno N. (2001) SECI, Ba and leadership: A unified model of dynamic knowledge creation, in Nonaka I. and Teece D. (eds.) *Managing Industrial Knowledge*. Sage Publications, London, 13–43.

OECD. (1999) *Boosting Innovation. The Cluster Approach*. OECD, Paris.

Passiante G. and Andriani P. (2000) Modelling the learning environment of the virtual knowledge networks: some empirical evidence. *International Journal of Innovation Management* **4**(1), 1–31.

Pekar P. (2001) Alliance enterprise strategies destroying firm boundaries, in McKnight L.W., Vaaler P.M. and Katz R.L. (eds.) *Creative Destruction*. MIT Press, Cambridge MA, 119–144.

Porter M.E. (1999) Clusters and competitions, in Porter M. (ed.) *On Competition*. Harvard University Press, London.

Quinn J.B. (1999) Strategic outsourcing: Leveraging knowledge capabilities. *Sloan Management Review*.

Rallet A. and Torre A. (1998) On geography and technology: proximity relations in localised innovation networks, in Steiner M. (ed.) *Clusters and Regional Specialisation*. J H L Dewhurst.

Rayport J.F. and Sviokla J.J. (1995) Exploiting the virtual value chain. *Harvard Business Review* November–December.

Reicheld F.F. and Schefter P. (2000) E-loyalty. *Harvard Business Review* July–August.

Romano A., Passiante G. and Elia V. (2001) New sources of clustering in the digital economy. *Journal of Small Business & Enterprise Development* **8**(1).

Sawhney M. and Parikh D. (2001) Where value lives in a networked world. *Harvard Business Review* January.

Sawhney M. and Prandelli E. (2000) Communities of creation: Managing distributed innovation in turbulent markets. *California Management Review* **42**(4), 25–55.

Saxenian A. (ed.) (1996) *Regional Advantage Culture and Competition in Silicon Valley and Route 128*. Harvard University Press, London.

Shapiro C. and Varian H.R. (1999) *Information Rules*. Harvard Business School Press, Boston.

Stacey R.D. (1995) The science of complexity: An alternative perspective for strategic change processes. *Strategic Management Journal* **16**.

Steiner M. (1998) The discrete charm of clusters: an introduction, in Steiner M. (ed.) *Clusters and Regional Specialisation*. J H L DEWHURST.

Storck J. and Hill P.A. (2000) Knowledge diffusion through strategic communities. *Sloan Management Review*.

Storper M. (1997) *The Regional World*. The Guildford Press, New York.

Tapscott D., Ticoll D. and Lowy A. (2000) *Digital Capital*. Nicholas Brealey Publishing, London.

Thompson J.N. (1994) *The Coevolutionary Process*. University of Chicago Press, Chicago.

Tidd J., Bessant J. and Pavitt K. (1997) *Managing Innovation–Integrating Technological, Market and Organisational Change*. John Wiley and Sons, Inc., New York.

Wenger E.C. and Snyder W.M. (2000) Communities of practice: the organisational frontier. *Harvard Business Review* January–February.

World Bank (1999) *The World Bank (1998–99)*. Oxford University Press Inc., New York.

Zenger T.R. and Hestley W.S. (1997) The disaggregation of corporations: Selective intervention, high-powered incentives and molecular units. *Organisation Science* **8**(3), 209–222.

Ziman J. (1991) A neural net model of innovation. *Science and Public Policy* **18**(1), 65–75.

Appendix: An overview of the case studies

Borland Software Corporation

Borland Software Corporation (Nasdaq NM: BORL) is a provider of technology that helps companies develop, deploy, and integrate software applications. Borland's vision is to enable a new digital world, where customers have the freedom to develop applications, deploy them on any platform, and integrate and manage them across the enterprise. Since 1983, Borland has been simplifying and speeding up the process of application development. As a pioneer in this space, Borland launched one of the first PC development environments, Turbo Pascal, which made the commercial development of PC applications possible. Through the years, Borland has refined its technology to meet the evolving demands of business environments. In 1996, Borland began expanding its offerings to serve a broader range of customers: the company launched a Java development environment with the award-winning Borland JBuilder, which is now the industry leader in the expansive Java development space. In 1997, the company acquired VisiGenic Software, enabling Borland to extend its application development expertise to enterprise application deployment. In the year 2001 Borland launched the Borland Kylix, a development environment for the Linux platform. Within six months, Kylix became the industry leader in this space. Today, Borland supports the major software architectures used to develop, deploy, integrate and manage enterprise e-business applications, including the Sun J2EE platform and the Microsoft.NET framework. Keeping pace with the rapid evolution of enterprise information technology, Borland has emerged as a key player in delivering development and deployment solutions for robust, standards-based Web Services and wireless applications.

DHL

DHL is a leader in worldwide services of shipping, tracking and tracing documents and packages.

DHL was founded in 1969 by Adrian Dalsey, Larry Hillblom and Robert Lynn (D, H, and L), who started a door-to-door express service between San Francisco and Honolulu. In 1971, they expanded to cover the Pacific Basin with services to the Philippines, and in 1972 services were activated in Japan, Hong Kong, Singapore, and Australia. In 1973 DHL expanded into Europe, and in 1976 began services to the Middle East and established a strong presence in Saudi Arabia.

In 1983 DHL brought its international air express service to Eastern European countries. In 1988 DHLNET, a high-speed global communications network used for shipment tracking, was established, and in 1990 DHL formed a strategic business alliance with Japan Airlines, Lufthansa, and the Nissho Iwai trading company.

In 1993 DHL announced a 4-year $1.25 billion worldwide capital spending program. The investment was to be made principally in handling systems, automation, facilities and communications and computer technology, and in 1995 it launched its site on the World Wide Web.

Elettrodata

Elettrodata is a producer of personal computers, founded in 1991, as a supplier of IT products, accessories and magnetic supports. Applying the technical know-how and marketing expertise acquired through the industrial application of technological solutions and on the global procurement markets, Elettrodata has achieved high growth rates by virtue of the quality of its products and the continuous re-investment of financial resources, becoming a reference brand in the Italian IT market.

In the past few years Elettrodata has established a firm foothold on the Italian market — opening nine locations including sales points, cash&carry shops and subsidiaries (Elettrodata Point) — and has diversified its business by setting up two new service companies in the Internet (ED.it) and technical support (Elettrodata Service) sectors.

The Gruppo ELETTRODATA concept was born in 1997, as the natural result of corporate expansion. The key strategies of Gruppo Elettrodata are diversification of products and sales channels, focus on quality, careful procurement policy. The production range including about 2,000 product ranges from complete systems branded Elettrodata to peripheral units from the leading global manufacturers, from components and accessories to software packages, up to telephony solutions.

In terms of distribution, the group has effectively developed both its geographical penetration through direct channels, and business-to-business distribution to specialised dealers. Today, they are committed to empowering the Internet and large distribution channels. Finally, in order to improve its procurement conditions, Gruppo Elettrodata has developed a procurement policy based on a limited number of suppliers, selected among the leading actors worldwide, establishing special relationships with them.

Fineco

Fineco products aim at retail clients, and include credit/debit cards, mortgages, personal loans, banking services and on-line securities trading. Fineco is also a brand leader in the area of small and medium sized businesses and of leasing and factoring. The "multi-channel" strategy means that the distribution of its products also has to be based on other Financial Institutions, and considers Internet a strategic channel to use for a push strategy. The strategic assets of Fineco are financial brokering (Fineco on line) and on-line bank transactions. Currently, Fineco provides a wide spectrum of services: from the continuous monitoring of current accounts to innovative bank teller machines, and offers attractively high interest rates.

Internet Bookshop Italia (IBS)

Internet Bookshop Italia is an on-line bookshop, realised by Informazioni Editoriali with Internet Bookshop, the biggest European on-line bookshop. Internet Bookshop Italia opened its virtual doors in June 1998, using the software systems provided by the Internet Bookshop of Oxford. Internet Bookshop Italia offers about 250,000 Italian titles that supplement the catalogue of 1,200,000 English titles. Subjects include History, Science fiction, Literature, Cookery, Romance, Science.

Wine

Wine, by eVineyard, is the largest on-line wine retailer in the United States. eVineyard, Inc., using wine.com as its branded storefront, is the nation's largest on-line wine retailer offering wine and related products. Formally launched in May 1999, eVineyard has emerged as the largest on-line wine retailer in the United States and the first truly national wine retailer legally serving wine buyers in 27 states and Japan, and selling gifts and accessories in 23 more. eVineyard, Inc.'s main sales and marketing channel is its wine.com branded website. In 2000, eVineyard customers and revenues grew by 1000%. As a result of its success, in April of 2001, eVineyard announced the acquisition of certain assets that were previously owned by leading rival Wine.com, Inc., including the wine.com and WineShopper.com URLs, and all of the goodwill and other assets of the wine.com service. In July of 2001, eVineyard became "wine.com by eVineyard," significantly enhancing its website and adopting "wine.com" as its primary brand. Thus, in just two years eVineyard has obtained its present preeminent position in on-line wine retailing.

With a vast selection of more than 5,000 wines, gift certificates, accessories, wine clubs, reviews and more, wine.com is the Web's premier wine retailer, offering a good selection and service. At the website, wine enthusiasts can:

- Find wines by brand, vintage, vineyard, rating, price, appellation or varietal;
- Review winemakers' notes;
- View wine labels;
- Learn which wines to serve with different foods;
- Subscribe to an e-mail newsletter;
- Join a wine club;
- Learn all about wine in the About Wine section.

	Fineco	Elettrodata	DHL	Wine	IBS	Borland
Strategic targets	• Maintenance of leadership • Cost lowering • Delivery time reductions • Market penetration • New products • Quality • Improvement of the hw/sw platform	• Trademark • Cost lowering • Delivery time reductions • Market penetration • Post-sales services • Increasing of the web traffic • New products • Quality • Improvement of the hw/sw platform	• Improvement of the hw/sw platform • Quality of services • Flexibility • Cost lowering	• Quality of services for clients • Lead time reductions • Cost lowering • Web-site improvement	• Quality of services for clients • Cost lowering	• Quality of services for clients • Lead time reductions
"Learning from market" processes	• Fineco community • Newsgroups • E-mail • Registration forms • Products selected • Browsing patterns	• Elettrodata points of sales • On-line communities • On-line forums • Call centres • E-mail • Market researches	• Participation at fears • Market researches • E-mail • Products shipped	• Discussion groups • Registration forms	• E-mail • On-line forums • Monitoring of preferences • Customer reviews	• Newsgroups
Strategic processes	• Strategic planning • New products development	• Strategic planning • Sales management	• Acquisition of information for monitoring clients • Information systems management	• Human Resources Management • Plannning of the plants	• Sales management • Marketing	• R&D • Information systems management

	Fineco	Elettrodata	DHL	Wine	IBS	Borland
	• Hw/sw platform management	• Hw/sw platform management				
Strategic alliances	• Intermobiliare Security SIM for research into corporations • Bridge for financial information • IlSole24Ore for financial news • Reuters for macroeconomic information • TIM for wireless services • Aereo Servizi Network Italia for outsourcing call centres • Lehmann Brothers for information about shares • Milano Finanza for analysis of the financial communities • Covered Warrants for financial transactions	• Asus Tek, Microston, Soltek, eCom for components • Microsoft, Norton for software • Elettrodata Points for franchising • Freedom Shops as retailers	• National Airlines (Lufthansa, Japan Airlines) • Insurance companies (AIG Europe) for insurance cover • National postal services • Customs for formalities	• Suppliers of wines • Shopper Comm Association for market penetration (advertising) • Affiliates • Logistics services (UPS, Manhattan Associates) • Bizrate.com for web-site evaluation	• SDA for logistics services • Messagerie Libri as a distribution channel • IBS Oxford for foreign books • Microsoft for software	• Sun and IBM for technological platforms

2

A Strategy to Sustain the Development Processes of a "Smart Community" in the Digital Economy

Valerio Elia

2.1. Introduction

The deep transformation the world economy has undergone in the last two decades has been characterised in several ways by different scholars. Following their own scientific backgrounds and specific academic disciplines, each scholar may refer to the new economic environment using such terms as Global Economy, Digital Economy, Learning Economy, Knowledge-based Economy, Alliance or Flexible Capitalism, etc. In this chapter, I focus mainly on how this transformation shapes the relationship between the innovation processes and the regional level of economic organisation in the more general framework of the economic development issue. This analysis drives the set-up of a new model, the Smart Community, to sustain the regional innovation processes in the new economic environment. With this intention, the three main objectives of this chapter are:

1. To recognise the special character of the new economic environment in which the evolution of Information and Communication Technology

(ICT) is a driving force enabling the transformation, hence the use of the term "Digital Economy" to define the new environment;
2. To set up the conceptual framework and the methodological tools needed to handle the regional innovation issue in the Digital Economy scenario. To this aim, based on the conclusions of Chapter 1 in this book about Virtual Clusters, I define the region-enterprise metaphor to borrow some guidelines useful for regions;
3. To set up the Smart Community model in terms of a strategy, for the firms and public and private institutions in a region, to meet the challenges of the Digital Economy.

In this work, the new regional foundations of economic performance, not only for the region as a whole, but also for the single firms located in that region are highlighted. In my view, two main factors shape the new role of regions in the economic system:

- The mobility, across national boundaries, of firms' created assets due to the globalisation processes which the world economy is undergoing;
- The increasing rate of learning processes inside organisations due to the emergence of a Knowledge-based Economy.

These factors require that regions provide immobile created assets to complement the mobile assets of firms. Among others, because learning is a collective and socially embedded process, regions should provide an institutional infrastructure in order to offer efficient support to learning processes within and among firms. Moreover, these factors represent a challenge to the management of firms because the choice of where to locate a branch, manufacturing plant, or home base is no longer a "definitive" choice, but depends on the new opportunities offered by a region. In this sense, the new competitive environment needs an interdisciplinary approach through the convergence of different traditional disciplines such as management and economic geography. Concomitantly, these factors represent a challenge for the stakeholders of a region because they increase the competition between regions in attracting and retaining firms.

This chapter proceeds in the following way. Section 2.2 introduces the Digital Economy environment adopting an evolutionary approach, and gives some stylised facts emerging from this approach: (1) the co-evolution of ICT and structures determines the emerging dynamics of the globalisation processes; (2) in comparison to the past, the structural adjustments in the Digital Economy radically shape the institutional set-up at each spatial and organisational level; (3) learning processes and organisational knowledge creation processes determine the innovation capabilities of firms and regions; and (4) in the Digital Economy the rate of change, and therefore the need for rapid learning, has increased.

In Section 2.3, I assume that the relationship between firms, regions, and innovation is a crucial issue in setting up a conceptual framework of regional innovation. The elements determining this relationship are: (1) the ability to create and deploy intangible assets (and more specifically knowledge assets); (2) the role of social capital, because it determines the dynamics of learning processes; and (3) the institutional environment, because in the evolutionary perspective, institutions and policies matter in shaping economic dynamics. In this section, the role of regional innovation networks in determining the economic performance of a region, and the network externalities concept to overcome the external economies concept are introduced.

In Section 2.4, I introduce the Smart Community model as consequence of the region-enterprise metaphor. This metaphor is based on the assumption that with regard to innovation performance and to knowledge generation, the dynamics governing an Internet-worked Enterprise and a regional innovation network are essentially the same. Thus, in setting a strategy to sustain the creation and maintenance of innovation networks one should take into account the new forms of organisation of firms, i.e. the Internet-worked Enterprise, and the enabling role of ICT in fostering networking. The main elements of the Smart Community model are: (1) an interactive communication infrastructure in the region to connect citizens, government institutions, territorial entities and firms; (2) the human capital to lead the innovation processes and to take advantage of the new media; and (3) several "vertical applications", in different industries, to stimulate the establishment of relationships, i.e. the development of the social capital. I also propose a methodology to develop the Smart Community in a real setting, following an issues-driven approach to develop applications and technology.

2.2. A New Techno-Economic Paradigm: The Digital Economy

Chapter 1 of this book provides strong empirical evidence, at firm and industry level, of the new economic environment characterised by the emergence of Virtual Clusters. It examines the transformations in technology, organisation and management that drive the search for new sources of value in the Digital Economy. Moreover, it proposes a rethinking of the role of the spatial dimension in the organisation of economic activities, introducing the shift from the national innovation system to the virtual innovation system.

In this section, I highlight some of the most striking facts about the Digital Economy at the macro level. I then use these facts to set up the conceptual framework of regional innovation in the Digital Economy and to build the Smart Community model.

The view taken in this chapter is that the transformation which the world economy is undergoing is enabled by the evolution of ICT. Hence, the term Digital Economy highlights the main force driving the transformation. In my view, the other definitions — Global Economy, Learning Economy, Knowledge-based Economy — refer to different subsets of the structural adjustments following the ICT revolution. In this view, according to Freeman and Perez (Freeman and Perez, 1988) and Lipsey (Lipsey, 1997, 1999) the Digital Economy represents a new techno-economic paradigm, whose key factors are the microprocessor and the Internet. The adoption of microprocessors and Internet by the overall economic system is producing a series of deep structural adjustments, the main of which, according to Lipsey (Lipsey, 1999), are:

- Changes in the organisation of firms. The old hierarchical firm gives way to a new flexible form of semi-independent groups linked laterally rather than vertically, as shown with the empirical evidence of Virtual Clusters given in Chapter 1 of this book;
- Changes in the economies of scale. In the manufacturing sectors, the introduction of ICT has drastically lowered the minimum efficient scale of production for many individual product lines; fixed costs are covered by producing many product lines so that economies of scope become more important than economies of scale;
- Shift to services. The proportion of the labour force in the manufacturing sector is steadily declining, whereas the proportion of the labour force in the services sector is steadily growing, to the point that the services sector is now the largest single sector by employment in all industrialised economies;
- Locational effects. As the Virtual Clusters phenomenology shows, the massive adoption of ICT allows production to be disintegrated into a series of independent operations, and moreover, it allows independent units to be co-ordinated in ways that were impossible in the past. These transformations, together with the improvements in transportation technologies, today allow component parts to be

produced anywhere in the world and to be shipped to arrive when and where they are needed.

The approach shown above is a specification of the more general evolutionary perspective (for good reviews see Nelson and Winter, 1982; Dosi *et al.*, 1988; Dosi, 2000), that offers a theoretical framework for dealing with issues of technological change, in this case changes in ICT, and economic dynamism. In this perspective, technology and institutions co-evolve through conflict-ridden processes (structural adjustments) to reach a "good match" in terms of economic performance. Here, by institutions I use the very broad definition given by Coriat and Dosi (1998):

- Formal organisations (firms, trade unions, universities, state agencies, etc.);
- Patterns of behaviours that are collectively shared (routines, social conventions, ethical codes, etc.);
- Negative norms and constraints (formal laws, moral prescriptions, etc.).

There are two main elements in the evolutionary approach which are useful for my analysis, and they stem from the observation of the "real life" of an economic system and from the above definition of institutions. They are:

- Technological change is endogenous in the economic system; therefore, institutions cannot avoid influencing the amount, location, and direction of technological change. The socio-institutional framework matters in shaping what firms do and how successfully they do it. As the main consequence of this statement, the role of government policies, including those related to education, competition, and redistribution, is crucial in influencing growth;
- The dynamical adjustments following a technological change are different in nature from those "allocative" mechanisms postulated by traditional theory. In the evolutionary approach, learning processes, selecting mechanisms, and institutional structures give continuous "disequilibrium" features to the economic system. Different combinations of learning modes, selection processes and institutions yield different environments with different performances and different evolutionary paths (Dosi and Orsenigo, 1988).

In what follows, I use these elements and the theoretical framework of the evolutionary approach to analyse the main effects of the ICT revolution at a global level.

2.2.1. *Globalisation processes in the Digital Economy*

Many of the structural adjustments enabled by the ICT revolution can be interpreted as an increased globalisation of the world's economies, hence the term Global Economy. According to Kobrin (1997), in the Global Economy, national markets are fused transnationally rather than linked across borders through flows of trade and investment as in the previous International Economy. The critical question is whether globalisation defines a change in degree or in kind — an extension of the modern international world economy into somewhat unfamiliar territory or a systemic transformation which entails both changes in quantity and quality, defining new structures and new modes of functioning.

In our view, the globalisation processes are a manifestation of the deep structural adjustments the world economy is undergoing. In this sense globalisation does have substantive meaning. We are in the midst of a qualitative transformation of the international world economy. The evolution of ICT and the changes in the structures, mentioned in the previous section, enable the shift from the International Economy to the Global Economy. Some of the main, interrelated, factors characterising this shift are:

- As Chapter 1 of this book has shown, ICT represents a means of integrating and co-ordinating extra-firm activities within Virtual Clusters. Moreover, ICT facilitates the integration of geographically-dispersed operations and allows networked coordination to replace ownership and hierarchy as the primary mode of control. The declining costs of communication, and of transportation, facilitate the establishment and monitoring of global production networks that enable firms to exploit international cost differentials through the fragmentation and relocation of production and global sourcing. Indeed, the ICT and transport revolutions have allowed production units to be de-linked from design units, and allowed parts that are manufactured throughout the world to be delivered to assembly plants when and where they are needed. By reducing the importance of transport and coordination costs, the ICT revolution has increased the importance of other variables such as international relative wages, the skills of the labour force, and the "friendliness" of government policy;
- The Global Economy is characterised by a strong international economic interdependence in goods and assets markets as highlighted, using statistical data, by a recent survey by the International Monetary

Fund (IMF, 2001). According to this survey, trade and financial market linkages represent important channels for transmitting disturbances internationally; among these linkages, financial linkages are stronger because of the growing rate of Foreign Direct Investments (FDI). Indeed, FDI can be regarded both as an international relocation of capital and as a basic means of establishing global production networks. As regards the drivers of FDI growth, there are: (a) the access to basic factors (capital, raw materials, etc.) from a foreign outlet, (b) the access to markets, and/or (c) the selective tap of particular skills, technologies, or learning experience (Dunning, 1997). In the new environment, more and more developed countries are in competition to attract high-quality FDI based on the last driver;

- Dramatic increases in the scale of technology — in its cost, risk, and complexity — have rendered even the largest national markets too small to be meaningful economic units (Kobrin, 1997). Otherwise, as previously remarked, the introduction of ICT has drastically lowered the minimum efficient scale of production, so it is the cost and risk of technology rather than the need for larger production runs that are the drivers of the transnational integration of markets. The increasing cost, risk, and complexity of technology are the most important (but obviously not the only) motivations behind the formation of strategic alliances, especially in some industries, such as automobiles, pharmaceuticals, biotechnology, aerospace, and information technology. In these industries, single firm globalisation no longer appears sufficient, as even the largest multinationals must cooperate to deal with the cost, risk, and complexity of technology. Cross-border strategic alliances, together with FDI, are the main drivers behind the geographical dispersion of economic activities;
- The emergence of networks as a basic mode of organisation of international economic transactions, hence the definition: "Network Economy". The Network Economy is conceived in terms of a complex web of transactions (Kobrin, 1997) based on alliances linking various parts of an organisation (for example an MNE) with others. As seen in Chapter 1 of this book about the phenomenology of Virtual Clusters, network forms of organisation represent a distinctive mode, in comparison to markets and hierarchies, of coordinating economic activity and economic organisation.

The most evident effect of the dynamics described above is a relentless erosion of the borders between individual national economies and a shift in the pattern of world development from a network of interacting national economies toward a single globally integrating economic system (Scott, 1998). This shift represents a challenge for national governments and they generate new governance mechanisms and patterns through re-allocations of power (Lipsey, 1997). Reallocations are tending to transfer some of the powers of national governments upwards to supranational bodies and others downwards to more local levels of government. These processes cause the emergence of a fourfold spatial hierarchy of economic and political relationships (Scott, 1998):

- The global level constituted by different networks of economic activity;
- The pluri-national level represented by a group of multination blocs, such as the EU, NAFTA, etc.;
- The national level focused on the classical sovereign state;
- The regional level that, according to Scott, is re-emerging as a modern articulation of economic organisation.

Despite the name, in the Global Economy it is the strong interdependence between the levels of this hierarchy that matters, rather than the global level alone.

2.2.2. *Knowledge and knowledge assets in the Digital Economy*

increasing relevance of knowledge and of knowledge assets in the Digital Economy is another effect of the structural adjustments which the world economy is undergoing as result of the ICT revolution. The empirical evidence about Virtual Clusters, as shown in Chapter 1 of this book, highlights the role of knowledge and of learning processes in the dynamics of cluster formation and growth. Others empirical evidences highlight that the intangible component of value far outweighs the value of their tangible assets. The growing relevance of knowledge and of knowledge assets in the economic system can be characterised in terms of a Knowledge-based Economy.

According to Senge, knowledge is the capacity for effective action and learning is the process whereby human beings build knowledge (2000). Otherwise, in Lundvall's view, knowledge is the most strategic resource and learning the most important process (1994). According to Lundvall, the focus should be on learning (hence the concept "Learning Economy") rather than on knowledge, because:

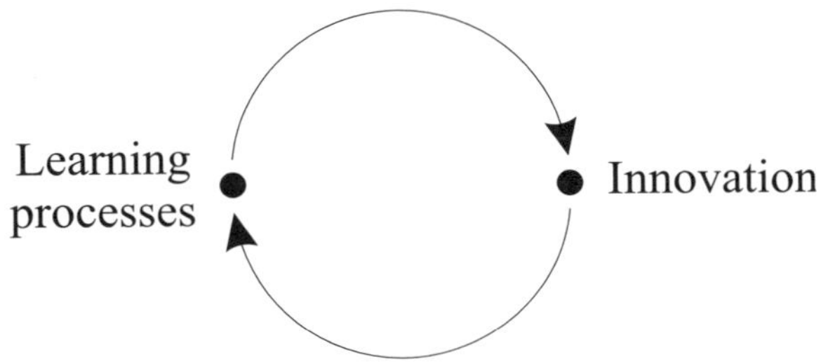

Fig. 2.1. The "wheel of learning" formed by learning processes and innovation according to an evolutionary approach.

- In the new economic environment the rate of change and therefore the need for rapid learning is increased;
- What matter for economic performance is not so much the knowledge possessed by agents and organisations at a certain time but rather the capability to learn (and forget).

In the evolutionary perspective, knowledge and learning processes are strictly related to innovation and, hence, to value creation. Learning processes are the first step in producing innovation, and, similarly, each innovation triggers new learning processes due to the need to deal with the change. In some sense, learning processes and innovation form a wheel like the one shown in Fig. 2.1, whose revolution speed is increasing in the Knowledge-based Economy.

In a previous work (Elia and Passiante, 2001), I highlighted the role of knowledge and learning processes in Virtual Clusters for wealth creation and how they represent a challenge to the conventional theories of how value is created. And indeed, the search for new sources of value creation inside firms requires new models for the creation and exploitation of intangible assets (and primarily of knowledge assets). Amongst others, I recall four different models explicitly treating knowledge and learning in management theories:

- The "learning organisation" model, mainly based on Senge's work (1990), in which the focus is on organisational learning as the process whereby organisations enhance their capacity for effective action;
- The "organisational knowledge creation" model, by Nonaka and Takeuchi (1995), in which organisational knowledge is created by tacit knowledge — explicit knowledge interaction processes at different ontological levels (individuals, groups, organisations);

- The "intellectual capital" models, introduced to take into account the growing discrepancy between market and book value in many knowledge-intensive companies. Among others, I recall the model of Seeman *et al.* (2000) in which the intangible assets of a firm are represented by three components: the "human capital", the "social capital", and the "structural capital", mutually interacting. In this view, knowledge management describes the management's efforts to enhance intellectual capital and to ensure that it generates superior business results;
- The "business innovation leadership" model (Romano, Elia and Passiante, 2001), that represents a tentative to build an holistic conceptual framework based on the previous models. In this framework, the pervasiveness of innovation in contemporary business fosters a shift in the unit of analysis of innovation, from product or process to business concept (Hamel, 2000). Business innovation leadership is defined as the capability to exploit the business opportunities deriving from innovations reconfiguring the strategic assets, and particularly the intellectual capital assets, of the organisation.

These models operate at firm level, and they represent new challenges for management. But how does the Knowledge-based Economy manifest itself at macro level, and how does it influence the spatial distribution of economic activities?

At global level, the search, creation, and deployment of knowledge assets are mainly transforming the economic environment for MNE activities and, particularly, they shape the FDI and strategic alliance activities. In this sense, there is a strong interdependence between these processes and the globalisation processes described previously. Some stylised facts on this interdependence are as follows:

- Increasingly firms are investing abroad, to protect or to increase their core competencies, accessing foreign created assets, i.e. technological capacity, information, human creativity, and markets. As remarked in the previous section, the selective tapping of particular skills, technologies, or learning experiences is overriding other factors as a driver of FDI growth in the Digital Economy;
- Some of these created assets are proprietary to particular foreign firms, others are more generally accessible to firms but immobile across geographical space;

- Concerning the created assets owned by foreign firms, the modalities of seeking created assets are mainly: FDI that takes the form of mergers and acquisition of foreign firms; cross-border strategic alliances especially within knowledge- and information-intensive sectors (Dunning and Wymbs, 1999);
- Immobile created assets are part of the technological, educational, and legal infrastructure that provides general inputs to firms.

For the purpose of this work, I concentrate mainly on geographically-bound knowledge assets. According to a survey by Dunning and Wymbs (1999), the geographical sourcing of knowledge assets are as follows:

- Access to professional, scientific and administrative personnel;
- Access to process and product technology and innovative capacity;
- Access to organisational structures and competencies and managerial capacity and competitiveness;
- The impact of consumer demand on the upgrading of product quality making for more product innovation;
- The effect of inter-firm rivalry;
- The linkages with related firms (industry agglomerative economies associated with clusters of suppliers, customers, competitors, etc.) and with universities and/or public and semi-public bodies.

2.3. The Conceptual Framework for Setting Up a Smart Community

A huge amount of literature deals with the issue of economic development — at regional, national, and global level — from different perspectives. Among these different perspectives I recall:

- The geography perspective based on the general theory of location in which "the central question is how and why different types of production occur at different levels of quantitative resolution in different places, and how any specific locational outcome affects the performative qualities of the economy" (Scott, 1998);
- The competitiveness perspective that addresses the role of location in competition, and is strongly influenced by Porter's theory (1990) of the competitiveness of nations, regions, and other geographic areas. Through his diamond metaphor, Porter depicts the nature of a location's business environment offered to firms, and how this environment

affects the firms' competitiveness. Moreover, Porter shows how the prevalence of clusters in economies, rather than isolated firms and industries, determines the nature of competition and the role of location in competitive advantage;

- The institutional economic perspective in which the pathway to economic growth and development is set by the character of established institutions and their ability to adjust to change in the most cost-effective way. In particular, the 1995 Nobel Prize winner Douglass North (1990) highlights how the role of institutions is to minimise the transaction costs through high-trust institutional framework and co-operation;
- The learning regions perspective based on the assumption that economic growth is dependent on innovation, and innovation, in turn, is dependent on learning processes; learning processes are generally believed to be connected with space, hence learning regions (for a review of learning regions see for example Boekema *et al.*, 2000). According to Florida, learning regions function as collectors and repositories of knowledge and ideas, and provide an underlying environment or infrastructure that facilitates the flow of knowledge, ideas and learning (2000);
- The systems of innovation perspective (essentially in the form of National Systems of Innovation, NSI) — based on an evolutionary approach to technological change — was developed mainly by Lundvall (1992), Nelson (1993) and Freeman (1995); see also an excellent review by Edquist (1997). A broader definition of NSI includes all parts and aspects of the economic structure and the institutional set-up affecting learning and involved in searching and exploring (Lundvall, 1992).

This list of perspectives is obviously an incomplete list of the several approaches to the economic development issue. From these different perspectives the following elements emerge:

a) They share the view that space, innovation, knowledge, learning, institutions and related issues matter, explicitly or implicitly, in determining the economic development of a region;
b) Some of these perspectives, like the learning regions and the systems of innovation perspectives, recognise, explicitly or implicitly, the distinguishing characters of the new economic environment, assigning a prominent role to the learning processes;
c) Some of these perspectives, like the institutional economics and the systems of innovation perspectives, assign a prominent role in their dynamics processes to the institutional set-up;

d) Each of these perspectives addresses a specific point of view, some of them have an interdisciplinary approach, and in my opinion, they stimulate the search for a coherent and holistic conceptual framework to deal with the issue of the regional development.

These elements, and particularly the last one, drive the set-up of a holistic conceptual framework to explore the issues of regional innovation in the Digital Economy. The two main objectives of this section are:

- To recognise the new role of regions in the Digital Economy, particularly in shaping innovation processes at firm and system level. This new role is deeply shaped by the global economic environment as described in the previous section in terms of Global Economy and Knowledge-based Economy;
- To introduce the "region-enterprise metaphor" to draw useful implications for the conceptual framework of a Smart Community.

2.3.1. *Regional innovation in the Digital Economy*

The organisation of production in the Digital Economy involves regions in a transformed way compared to their role in previous modes of economic co-ordination.

A lot of literature (see, among others de Vet, 1993; Ohmae, 1995; Florida, 1995; Cooke, 1995; Sabel, 1995; Storper and Scott, 1995; Cooke and Morgan, 1998; Scott, 1998) make the case that the regional level of economic organisation has become more important in the Digital Economy which emerged from the 1980s and 1990s restructuring process. In reality, as remarked in the previous section, the Global Economy has a fourfold hierarchy — global, pluri-national, national, and regional — that matters in shaping the organisation of economic activities, rather than the global level alone. Indeed, localised processes of growth and development are emphasised by globalisation, and the world economy is characterised by strongly interconnected regional economies.

At the heart of the regional foundations of economic performance is the concept of "agglomeration" as the source of competitive advantage for regions. Indeed, the geography of industrial development shows a pattern of local clusters of interconnected firms, specialised suppliers, service providers, firms in related industries, and institutions, characterised by two main attributes:

- Externalities, i.e. any occurrence or activity that lies outside the range of control of the individual firm, but that then has definite effects on the firm's internal production function;
- Geographical proximity resulting from the co-location of firms in a region.

In reality, externalities and geographical proximity are interrelated concepts, because many externalities derive from proximity. For example, the externality effects from economies based on proximity through agglomeration confer advantages in terms of transaction costs. In an evolutionary perspective of regional development (Cooke and Morgan, 1998), proximity in cluster dynamics also matters in determining:

- Path-dependence, cumulative causation processes, and increasing returns effects, that generate developmental trajectories through processes of evolution and adjustments structured by the phenomenon of agglomeration economies (Scott, 1998). Path-dependence and increasing returns effects play a role in the growth processes of clusters through self-reinforcing phenomena of industrial concentrations;
- The creation of a "milieu" (Maillat, 1995) or institutional infrastructure that, as seen in previous sections of this chapter, matters in shaping economic organisation. In cluster theories, face-to-face interactions enabled by proximity contribute to the sharing of common norms, rules, and routines.

The Digital Economy is boosting a shift in the nature of agglomeration economies: from static towards dynamic efficiencies based on innovation and rate of learning. Traditional agglomeration theories stress the relevance, for locational advantages, of cost minimisation due to the proximity to natural resources, related firms, customers, or skilled pool of labour. These advantages have been superseded by the globalisation of markets, technology, and supply sources, and by lower transportation and communication costs. Today, economies of agglomeration are based on the availability and quality of location-specific created assets that complement the mobile created-assets of firms; among these assets there are the institutions that foster entrepreneurship and knowledge accumulation (Dunning, 1997).

How must the institutions of a region be shaped in order to be effective in complementing the mobile created-assets of firms? And then, effective in attracting and retaining firms inside that region? Because agglomeration economies in the Digital Economy are based on innovation and rate of learning, we have to delve deeper into the nature of these processes so as to understand what kind of externalities regions must generate for firms.

Since learning is an interactive and socially embedded process, the institutions, as defined previously in this chapter, shape the learning efficiency of an organisation. Indeed, organisational learning critically depends on sharing knowledge (tacit, codified or both) but it also depends on trust and commitment among the members of the organisation itself, or among members of different organisations. Without a minimum of social cohesion, the capability to learn to master new technologies and new and more flexible forms of organisation will be weak (Lundvall, 1999). In cluster theories, trust is generated by face-to-face interactions, by the sharing of common rules and routines, or by belonging to the same community, i.e. trust is fostered by the institutions of a region.

The role of trust relationships and of collective value systems in affecting regional development emerges in the "social capital" concept popularised by Putnam (1993). Following Putnam, social capital is the result of the common history (along the centuries) shared by the members of a local or regional community. The institutions and organisations of this community, with their social relationships, form "networks of civic engagement", that facilitate the activities of politics, production, and exchange. In such communities, people know one another and they meet frequently in non-work related organisations and activities. Business relationships are embedded in community and family structures, and these are structures that reinforce trust.

But if it is the past that establishes a certain pathway for the creation of social capital networks, and if, by definition, the past is basically fixed, how then can social networks be created? In this sense the concept of social capital is very limited for framing policy choices. On the other hand, if Putnam's model explains regional development in a country like Italy, it does not fit the experiences of other regions like Silicon Valley: "the sources of technological dynamism in Silicon Valley can be described in many ways, but there is little truth in the idea of Silicon Valley as a community of dense civic engagement" (Cohen and Fields, 1999). In Silicon Valley, social capital can be understood in terms of the collaborative partnerships that emerged in the region fostered by economic and institutional actors in search of objectives related specifically to innovation and competitiveness. Indeed, this interactive model of innovation carries radical implications not just for firms but also for a wide array of public and private institutions. This model "underscores the importance of cooperation between firms and institutions and, thus, the role played by links and networks involving different organisations" (OECD, 1992). The innovation networks (Rallet and Torre, 1998; Cohen and Fields, 1999) link firms, banks, research organisations, universities, government, and any kind of public and private institutions. In these innovation networks, the linkages among the actors are knowledge-based ones (Nonaka and Takeuchi, 1995; Rallet and Torre, 1998) which enable learning processes in an organisation or in a region.

According to Cohen and Fields what these innovation networks in Silicon Valley share with the networks of civic engagement is simply and only a common network-like structure. The fact that economic and institutional actors in Silicon Valley fostered the creation of innovation networks provides encouragement for efforts to create an innovative milieu in other regions.

In the next section, I introduce a network approach to regional development and I propose a shift from external economies to network externalities.

2.3.2. *Networks of innovation and network externalities*

The role of networks as new forms of economic organisation differing from markets and hierarchies has been emphasized in Section 2.2 of this chapter. Hierarchies have limited learning abilities and markets have limited capacities to process information effectively. Networks and alliances are ways to counter these failures, ways to combine the benefit of being large and small at the same time (Acs, de la Mothe and Paquet, 2000).

As the empirical evidence of Chapter 1 of this book shows, networks are the organisational forms on which Virtual Clusters are based. The concept of network highlights the role of cognitive and social relationships among the firms in Virtual Clusters and between firms and other organisations in a region. Moreover, it highlights the role of a value network in which an organisation needs to collaborate not only on a peer level, department level, or organisation level but also needs to see and collaborate holistically at a system level. The only way to achieve this result is by creating a network of relationships between every part of its "business ecosystem". At a time when the value chain concept is giving way to value networks, the emphasis needs to shift from minimizing transaction costs incurred by individual firms to maximizing transactional value created by networks of firms.

In a value network, due to the complex web of relationships among the participants, the boundaries of a firm are not well defined. If one focuses on an intangible asset such as social capital — i.e. the overall set of relationships that make organisations work effectively (Prusak and Cohen, 2001) — which part of the social capital is "internal" to the firm and which part is "external"? In this case, what is the meaning of the "external economies" defined previously in this section?

According to Saxenian, the concept of external economies is based on the view of the firm as an atomistic unit of production with clearly defined boundaries (2000). In cluster theories, external economies help to explain the advantages that are derived from the spatial clustering of economic activities. Saxenian, based on the different evolution and performances of Silicon Valley and Route 128, adopts a network approach to regions, in which the shift is from external economies to network externalities.

Here, I use the term "network" to mean a set of actors and the linkages among them. Learning and knowledge creation are interactive and collective processes that involve the creation of some kind of networks among individuals and organisations. According to Saxenian, Silicon Valley has a regional network-based industrial system and dense social networks that enable learning and mutual adjustments. In contrast, Route 128 is dominated by large vertical corporations, that internalise many productive activities within a hierarchic structure. The network approach helps to explain the differences in industrial adaptation of these two regions, under the same external forces. As seen in the previous section, Silicon Valley highlights the existence of "regional innovation networks" in which the functional boundaries within firms are porous, as are the boundaries between firms, and between firms and local institutions such as trade associations, universities, etc. (Saxenian, 2000). In these networks, firms compete but also learn from each other through informal communication and collaborative processes.

The advantages of networks over other forms of organisation are the "network externalities" features (Shapiro and Varian, 1999): they refer to the attraction power of large networks compared with small ones. Network externalities are what lie behind *Metcalfe's law*: if there are n actors in a network then the total value of the network is proportional to $n \times (n-1) = n^2 - n$. So a tenfold increase in the size of the network leads to a hundredfold increase in its value. For example, communication or transportation networks have this feature: the more people or destinations they can reach, the more valuable networks become.

Despite Metcalfe's law, the extension of an innovation network cannot grow indefinitely. Indeed, in this case, the geographical extension is limited by the nature of the relationships involved in the linkages of these networks. Informal communication, trust, and commitment are strongly dependent on face-to-face interactions. Electronic communication enables some kinds of collaborative processes, but it cannot totally replace the networks of social relationships enabled by geographical proximity. In this sense, the regional extension of innovation networks is the result of two balancing forces:

- Network externalities foster the growth of the geographical extension of networks through the involvement of new actors and the establishment of new linkages;
- Social capital creation and maintenance limit the size of networks into a homogeneous institutional system. Networks crystallize around a unifying purpose, mobilizing independent members through voluntary links, around multiple leaders in integrated levels of overlapping and superimposed webs of solidarity (Acs, de la Mothe and Paquet, 2000).

The creation and maintenance of innovation networks are a concern of technology policy. Public actors try to develop regional technology policies directed towards supporting collective processes of research and innovation. In the evolutionary perspective, the goal of these policies would not be to eliminate market imperfections and lead back to equilibrium, but to move the system deliberately away from equilibrium (Saviotti, 2001) through discontinuous or qualitative changes that will lead to the formation of a completely different network, with new actors and new linkages. Indeed, as the new system develops, its connectivity (the density of existing links) and the role of different actors will undergo systematic variations, with the emergence of a new set of institutional infrastructures. But more than one institutional configuration can lead to an equivalent outcome (Saviotti, 2001), thus the problem would become how to create the right network.

2.3.3. *Implications for technology policy at regional level*

What follows from previous sections is that policy implications should be focused on the search for an enabling strategy for innovation network creation and maintenance. On the basis of the conceptual framework of this chapter, several assumptions can be made:

- The distinction between industrial policy and technology policy is becoming less and less pronounced. Technology policy is becoming the main part of any industrial policy strategy, and it should influence all the factors that promote technological innovation, primarily the learning processes. According to Lundvall, in attempting to further technological innovation, one must start with the existing knowledge base, in the given institutional context (1999);
- The emergence of the Knowledge-based Economy should lead to a shift in the technology policy: from supply-oriented technology policy to innovation policy seeking a better balance with demand-oriented factors (Vonortas and Tolnay, 2001). A more balanced supply-side/demand-side technology policy is more appropriate for an innovation-driven economy focused on the distribution of knowledge, skill, entrepreneurship, and the facilitation of new forms of collaborating between firms, universities, and the government. This view highlights the network dimension of innovation processes and the needs for the establishment of innovation networks;
- Innovation networks must be analysed as a whole. A multiplicity of actors and relationships among them determine the performance of the system. In these networks, institutions enable and facilitate innovation and learning processes. Missing linkages between firms and other public and private organisations — universities, local government institutions, etc. — can produce a mismatch among the different components of the system;
- Networks are consensus and inducement-oriented systems (Acs, de la Mothe and Paquet, 2000). This logic does not abolish power but it means that power and leadership are distributed and there is a "voluntary" adherence to norms. The dynamic of an innovation network is generated by trust and self-reinforcing mechanisms that enhance the social capital and generate increasing returns. Trust is the result of the complex web of relationships of the economic and institutional actors in pursuit of explicitly competitive aims, as in Silicon Valley (Cohen and Fields, 1999);
- Human capital is a critical component in an innovation network. Distributed leadership, as in innovation networks, calls for a collective capacity to start up and to sustain significant changes (Romano, Elia, and Passiante, 2001). In an innovation network, the role of human capital is to exploit the opportunities deriving from innovations reconfiguring the knowledge assets of the network itself;

- The impact of policy strategy on the social cohesion of the economy as a whole should be taken into account. The speed-up of change requires the capability to learn and adjust faster than in the past. Learning and adjustment processes stress the norms, the rules, and more generally the institutions of the system.

2.4. The "Smart Community" Model

The conceptual framework described in the previous sections represents the guideline for setting up the Smart Community model. To this aim, some stylised facts emerge from the conceptual framework and from the Digital Economy scenario:

- The new competitive environment is the result of the evolution of information and communication technology, and to some extent also of transportation technology;
- The Virtual Clusters phenomenology described in Chapter 1 of this book is the main evidence, at firm level, of the Digital Economy environment. It highlights the critical role of organisational learning processes and the growing relevance of intangible compared to tangible assets;
- The co-evolution of ICT changes and institutions is also producing a transformation of economic organisation at regional, national and global level;
- The regional foundations of economic organisation are no longer related to the external economies, as in the Marshallian view, but are related to the network externalities of innovation networks;
- In this new view, what is critical is the capability to create and maintain innovation networks between individuals, firms, universities, government institutions, public and private organisations of a region;
- Social capital is the "glue" of these networks. It is the result of the relationships between the actors of an innovation network that share common objectives in terms of innovation and economic performance. The social capital of a region critically affects the efficiency of the learning processes at intra- and inter-organisation level;
- Social capital and any other created knowledge asset that can help firms to exploit their own created assets are the main sources of competitive advantages for regions.

In the new perspective of regional innovation networks, as highlighted in the points above, geographical proximity determines:

1. The relationships between the internal organisation of firms and their connections to one another and to the social structures and institutions of their particular localities;
2. The level of trust and commitment among the actors of a region in sharing and producing knowledge, and thus the efficiency of learning processes.

As seen in the case of Silicon Valley and Route 128, it is not necessarily geographical proximity alone that produces the right network, so the problem is which other enabling factors can produce the right one. To this aim, I propose the region-enterprise metaphor as the new paradigm to enable regional innovation networks in the Digital Economy.

2.4.1. *The region-enterprise metaphor*

As seen in Chapter 1 of this book, the network organisation of Virtual Clusters, based on adoption of ICT, allows more efficient learning and adjustment processes in a fast changing environment. The typical unit of a Virtual Cluster is the Internet-worked Enterprise (IE) whose main features, as emerging from Chapter 1 of this book, are:

- The adoption of ICT, and specifically of e-Business technologies, to integrate and to coordinate the activities and the core competencies of the actors in the supply chain;
- The flux of information, available on the digital infrastructure, generated by the several activities and processes that take place in the supply chain;
- The community-like environment, enabled by this flux of information and by cooperative practices, among the several actors of the supply chain;
- The new role of customers, who together with the other actors in the supply chain create knowledge and innovation through learning-by-interacting processes;
- The role of intangible assets compared to the tangible assets as a result of the increasing relevance of organisational learning and innovation processes for the competitive advantage of firms.

Strategies aiming to sustain the creation and maintenance of innovation networks should take into account these new forms of organisation and the enabling role of ICT in fostering networking. Indeed, the network form of the IE facilitates the creation of linkages with the actors in its supply chain, through informal communications and collaborative practices.

In this sense, regarding learning and innovation processes, in the Digital Economy regional innovation networks behave like IEs:

- Like in an IE, in a regional innovation network several actors form a community sharing information and common practices. Also, as the Silicon Valley innovation network shows, in such networks actors share common objectives in terms of innovation and economic performance;
- Like in an IE, in regions the role of the intangible created-assets complementing firms' assets is increasing. As previously remarked, increasingly firms are investing abroad, to protect or to increase their-core competencies, accessing foreign created assets, i.e. technological capacity, information, human creativity, and learning experience;
- Like in an IE, in a regional innovation network the network form of organisation increases responsiveness and adjustment to the fast changing world.

Obviously, this metaphor needs some caveats. An IE has a stronger and more formalized coherence than a region. An IE is an "economic" agent, i.e. an agent whose behaviour is mainly driven by economic performance, while regional innovation networks also include "non-economic" agents, i.e. agents whose behaviour is mainly driven by social and cultural performances. In this sense a region has a different kind of goal-setting procedure. But, with regard to innovation performance and to knowledge generation, the dynamics governing IE and regional innovation networks are essentially the same. A basic element here is the development of channels of communication to transmit knowledge to other individuals and organisations. The role of ICT as enabler of learning and knowledge creation processes, thus determining the innovation performance of an IE as highlighted in Chapter 1 of this book, suggests the use of ICT as an enabler of regional innovation networks. In this sense, the adoption of ICT in a region can produce the discontinuous or qualitative changes that will lead to the formation of a completely different network, with new actors and new linkages, as remarked in a previous section of this chapter.

The previous statements drive the set-up of the Smart Community model, which shows how a region can benefit from the digital connectivity, and how the digital connectivity can be the enabler of "relational" connectivity and of social capital generation.

2.4.2. *Towards a Smart Community*

A Smart Community is a geographical area ranging in size from a neighbourhood to a multi-county region whose citizens, organisations and government institutions "use" ICT to strategically position that region for economic, social and technological development. The extent to which communities can take advantage of ICT is directly affected by its degree of "connectivity". Connectivity refers to the access to the global digital network that includes satellites, cable TV, new wireless systems, mobile telephones and the Internet.

But according to Lundvall (1999), investment in technology which neglects user competence, social needs and the need for organisational renewal can do more harm than good in the context of the Learning Economy. As seen in previous sections, the evolutionary approach privileges the co-evolution of technology change and institutions. Technology is an instrument and not a goal in itself. As with other instruments, the positive impact will depend on the institutional framework around it. In this sense, it is the "institutional-technological infrastructure", rather than the "technological infrastructure" that is to be regarded as the correct concept (Lambooy, 2000).

According to this view the transformation of a traditional community in a Smart Community can be obtained by stimulating the creation of the following (see Fig. 2.2):

1. An interactive communication infrastructure in the region to connect citizens, government institutions, territorial entities and firms;
2. The human capital to lead the innovation processes and to take advantage of the new media;
3. Several "vertical applications", in different industries, to stimulate the establishment of relationships, i.e. the development of social capital.

As Fig. 2.2 shows, points 1 and 2 refer to investments in people and technology, specifically ICT, which should trigger changes and innovation in every domain of the economic and social life of a region. Investments in people and technology are the main factors in developing a technology policy which recognises the growing relevance of learning processes in the economic development of a region. Learning is essentially an individual process, hence the relevance of human capital creation. But, as remarked in previous sections of this chapter, it develops in an interactive configuration that needs several channels of communication, thus the relevance of a digital infrastructure. Here, the adoption of ICT, and particularly of a

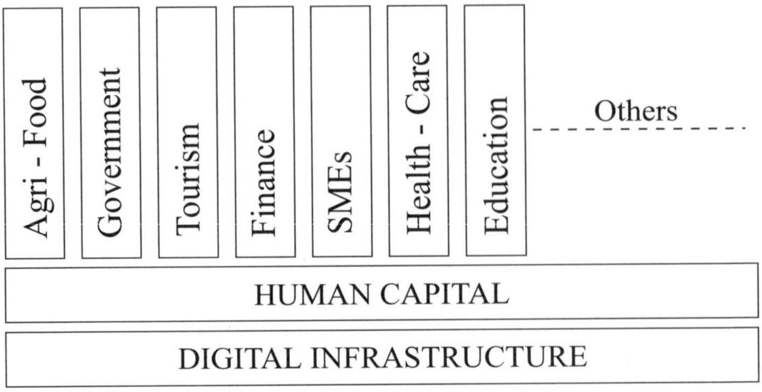

Fig. 2.2. The logical architecture of the Smart Community model.

large-bandwidth digital network, has two main objectives:

1. To support the learning and knowledge creation processes at intra- and inter-organisational level. As seen in Chapter 1 of this book (see also Romano, Passiante and Elia, 2001) knowledge creation processes involving tacit knowledge can be triggered by interactive tools. These tools, such as videoconferencing systems providing visual and real-time communication, have suffered until now from limited bandwidth availability;
2. To facilitate the coordination mechanisms among different actors in a region. Digital networks allow implicit coordination through the sharing of information resources, the access to common data repositories, the use of Decision Support Systems (DSS) operating in a network configuration. Such systems can offer different views of the same event depending on the profile of the user: firm, public institution, local government institution, etc.

The relevance of coordination mechanisms emerges in the discussion about social capital creation and maintenance. As seen previously, social capital is the result of the collaborative partnerships that emerged in the region fostered by economic and institutional actors in search of objectives related specifically to innovation and competitiveness. In this sense, social capital creation and maintenance need some objectives to be shared as well as some coordination mechanisms to support collaborative partnerships. Implicit coordination mechanisms are fostered by the adoption of ICT; the set-up of projects, i.e. the vertical applications in Fig. 2.2, gives innovation and competitiveness objectives.

The role of vertical applications needs some explanation. Their objective is to innovate several sectors of the economic and social life of the region, involving different actors. Innovation in the government sector means, for example, new services for firms and citizens, new ways to deliver these services, new laws, etc. This objective can be reached through collaborative partnerships between local government officials, entrepreneurs or firm personnel, groups of citizens, and any other actor inside or outside the region. At industry level, applications to reach innovation and competitiveness, mean for example new tools and services for the supply chain, or new local development agencies, the creation of linkages with universities or research centres. In the next chapters of this book, some examples will be given of vertical applications in the agri-food and tourism industries.

2.4.3. *The action plan to develop a Smart Community*

The Smart Community model shown in the previous section needs an action plan to be implemented in a real setting. In this section, I propose a methodology to develop the Smart Community in a region through the involvement of the entire community of that region. This methodology is the result of a real setting implementation in Lanark County, Ontario, Canada. In the next sections, a description of the "eSalento project" will be given as an example of a running Smart Community implementation in the Salento region, Italy.

I propose the creation of a planning group made up of government officials, entrepreneurs and firm personnel, university and research centre personnel, etc. The planning group follows an issues-driven approach to develop applications and technology. This approach addresses four areas in sequence: issues, applications, technology and products. The analysis of the region's issues, by following this sequence, produces network products that are solutions to specific issues targeted by the community. More specifically, the planning group should adopt a virtuous circle with several iterated steps like in Fig. 2.3.

The circle in Fig. 2.3 indicates that each step must be revisited, with a feedback mechanism, on the basis of the results of the successive steps. As Fig. 2.3 shows, the issues-driven approach promotes an advanced telecommunications infrastructure through the development of sophisticated network applications. The futures of the applications developed by the Smart Community are (1) based on the specific needs of the community; (2) can attract financial support; and (3) can be quickly moved

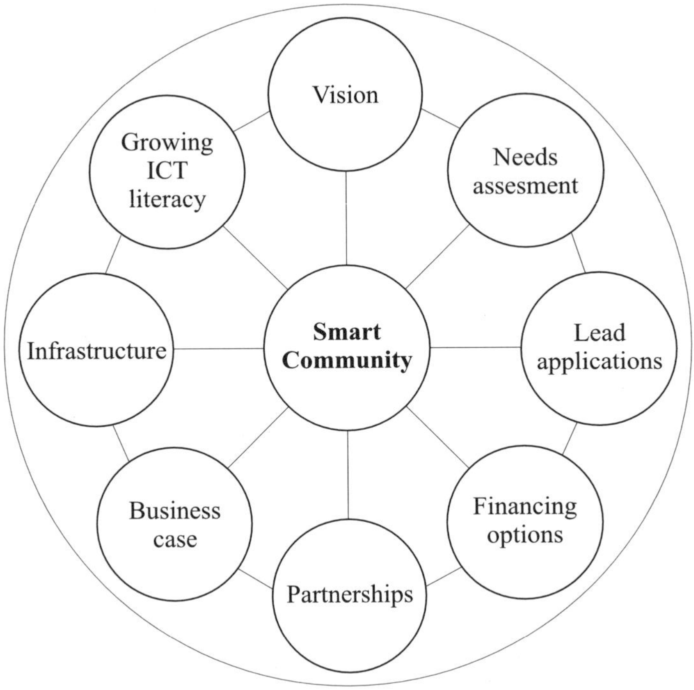

Fig. 2.3. The action plan to develop a Smart Community.

to financial self-sufficiency. Furthermore, the applications have an impact that goes beyond a specific constituency to affect the entire community. The meaning of each step is:

- Growing ICT literacy. This is a crucial step in the action plan because it refers to training and communication about the benefits of using ICT. As seen previously, technology alone cannot produce benefits; it needs people to use it. Furthermore, this step should generate awareness about the use of ICT so as to participate actively in the successive steps in the circle;
- Vision. This refers to setting up a vision of the Smart Community project and to identifying some general objectives. The vision and the objectives should be revisited during the development of the project because of the new ideas and new needs emerging after the involvement of the entire community in the project itself;
- Needs assessment. This refers to the identification of the factors of weakness and strength of the region, and to the identification of the

issues to be addressed through the Smart Community model. With the "growing ICT literacy" step the actors of the region should provide conscious ICT-related needs. This step drives the following one in setting the vertical applications;

- Lead applications. This refers to the identification of the vertical applications on the basis of the issues to be addressed. As seen in the previous section, vertical applications should promote innovation and competitiveness objectives to foster social capital creation and maintenance;
- Financing options. To develop the applications and the infrastructure of the project one needs to attract financial support. Actors involved in the project, government institutions and venture capital investors can provide this financial support;
- Partnerships. For each application one needs a group of actors to be involved in the development of the application itself. This step is strongly connected to the previous one because some of the partners in the project may also be financial investors in the project;
- Business case. A business case should be prepared for each application and for the infrastructure to build. The business case can demonstrate the financial self-sufficiency of each application;
- Infrastructure. The design and the futures of the digital infrastructure should be the result of the aggregation of the several applications identified. Indeed, the aggregation of the applications provides the bandwidth needed for the project.

Among the various types of applications that can be implemented in a Smart Community, as emerging from the Lanark County implementation, are:

- Teleradiology/Remote Diagnostics: through the use of remote medical diagnostics, communities in territorially peripheral areas can have access to the best, highly specialized medical care that may not otherwise be readily available — such as timely medical diagnosis and access to specialists outside the local community;
- Integrated Patient Records: this application benefits the community by providing timely and up-to-date access to the health records of patients served by separate healthcare service facilities;
- Teleworking: with the appropriate telecommunications infrastructure and services in place, communities can benefit from new employment opportunities. Workers can be employed either from home-based offices or Teleworking Centres;

- Municipal Amalgamation and Land Registry: through this network application, there is an opportunity to use telecommunications to deliver information and services to the public and to better manage the centralization of certain public administration functions;
- Opportunity Network: to attract investment from businesses interested in locating or expanding in the community, there is a need to present a detailed view of the community's strengths, advantages and capabilities to the widest possible audience of companies and investment intermediaries. A sophisticated telecommunications capability is necessary to deliver detailed economic information to businesses throughout the community and around the world;
- Geographic Information Systems (GIS): as local governments and municipal service providers are being required to do more with fewer resources, many are looking to information technology to improve the way they do business. GIS applications have proven effective as a governmental support system;
- Web-based Learning: citizens living in territorially peripheral areas face substantial barriers in accessing higher education and training due to distance and costs. The delivery of shared courseware between urban and rural centres using eLearning techniques reduces costs and provides users with courseware that would otherwise be unavailable.

2.5. Conclusions

Starting from the stylised facts on the new competitive environment enabled by the evolution of ICT, and by the several approaches to the economic development of a region, I tried to develop an integrated and holistic approach to the regional foundations of economic organisation in the Digital Economy. This approach includes the following elements:

- The co-evolution of ICT changes and institutions is enabling a new economic and social environment for firms and for regional competition. The new environment manifests itself as the Global Knowledge-based Economy;
- In the Global Knowledge-based Economy, learning and knowledge creation processes are crucial at both firm and regional level. These processes involve a social dimension that can be represented through the concept of social capital;

- The new regional foundations of economic organisation are strongly connected to the creation of an economic and social environment to enable and to support learning processes inside firms and organisations;
- Regional innovation networks and network externalities are the main concepts involved in these new foundations. More in particular, network externalities and social capital represent the balancing forces determining the spatial extension of an innovation network;
- The Smart Community model is the conceptual framework used to enable and to sustain a regional innovation network through the adoption and use of ICT. In this case, ICT acts as a trigger for a discontinuous and qualitative change that can produce new institutional arrangements and configurations.

References

Acs Z.J., de la Mothe J. and Paquet G. (2000) Regional Innovation: In Search of an Enabling Strategy, in Acs Z.J. (ed.) *Regional Innovation, Knowledge and Global Change*. Pinter, London, 37–49.

Boekema F. *et al.* (2000) *Knowledge, Innovation and Economic Growth: The Theory and Practice of Learning Regions*. Edward Elgar Publishing, Massachusetts.

Cohen S.S. and Fields G. (1999) Social capital and capital gains in Silicon Valley. *California Management Review* **41**(2), 108–130.

Cooke P. (ed.) (1995) *The Rise of the Rustbelt*. University College London Press, London.

Cooke P. and Morgan K. (1998) *The Associational Economy: Firms, Regions, and Innovations*. Oxford University Press, New York.

de Vet J. (1993) Globalisation and local and regional competitiveness. *STI Review* **13**, 89–121.

Coriat B. and Dosi G. (1998) The institutional embeddedness of economic change: An appraisal of the "evolutionary" and "regulationist" research programmes, in Nielsen K. and Johnson B. (eds.) *Institutions and Economic Change: New Perspectives on Markets, Firms and Technology*. Edward Elgar Publishing, Massachusetts, 3–32.

Dosi G. and Orsenigo L. (1988) Coordination and transformation: An overview of structures, behaviours and change in evolutionary environments, in Dosi G. *et al.* (eds.) *Technical Change and Economic Theory*. Pinter, London, 13–37.

Dosi G. *et al.* (eds.) (1988) *Technical Change and Economic Theory*. Pinter, London.

Dosi G. (2000) *Innovation, Organization and Economic Dynamics*. Edward Elgar, Cheltenham.

Dunning J.H. and Wymbs C. (1999) The geographical sourcing of technology-based assets by multinational enterprises, in Archibugi D. *et al.* (eds.) *Innovation Policy in a Global Economy*. Cambridge University Press, Cambridge.

Dunning J.H. (1997) A business analytic approach to governments and globalization, in Dunning J.H. (ed.) *Governments, Globalization and International Business*. Oxford University Press, London, 114–131.

Edquist C. (ed.) (1997) *Systems of Innovation: Technologies, Institutions and Organizations*. Pinter, London.

Elia V. and Passiante G. (2001) Setting the framework for business innovation leadership, in Romano A., Elia V. and Passiante G. (eds.) *Creating Business Innovation Leadership. An Ongoing Experiment: The e-Business Management School at ISUFI*. Edizioni Scientifiche Italiane, Napoli, 41–79.

Florida R. (1995) The industrial transformation of the great lakes region, in Cooke P. (ed.) *The Rise of the Rustbelt*. University College London Press, London, 162–176.

Florida R. (2000) The learning region, in Acs Z.J. (ed.) *Regional Innovation, Knowledge and Global Change*. Pinter, London, 231–239.

Freeman C. and Perez C. (1988) Structural crises of adjustment: Business cycles and investment behaviour, in Dosi G. *et al.* (eds.) *Technical Change and Economic Theory*. Pinter, London, 38–66.

Freeman C. (1995) *Cambridge Journal of Economics* **19**, 5–24.

Freeman C. (1997) *Innovation Systems: City-State, National, Continental and Sub-National*. SPRU, University of Sussex, December.

Hamel G. (2000) *Leading the Revolution*. Harvard Business School Press, Boston.

IMF (ed.) (2001) *World Economic Outlook: The Information Technology Revolution*. IMF, Washington, October.

Kobrin S.J. (1997) The architecture of globalization: State sovereignty in a networked global economy, in Dunning J.H. (ed.) *Governments, Globalization and International Business*. Oxford University Press, London, 146–171.

Lambooy J. (2000) Learning and agglomeration economies: Adapting to differentiating economic structures, in Boekema F. *et al. Knowledge,*

Innovation and Economic Growth: The Theory and Practice of Learning Regions. Edward Elgar Publishing, Massachusetts.

Lipsey R.G. (1997) Globalization and national government policies: An economist's view, in Dunning J.H. (ed.) *Governments, Globalization and International Business*. Oxford University Press, London, 73–113.

Lipsey R.G. (1999) Sources of continued long-run economic dynamism in the 21st century, in OECD (ed.) *The Future of the Global Economy. Towards a Long Boom?* OECD Press, 33–76.

Lundvall B.A. (ed.) (1992) *National Systems of Innovation: Towards a Theory of Innovation and Interactive Learning*. Pinter, London.

Lundvall B.A. (1994) The learning economy: Challenges to economic theory and policy. Paper to the EAEPE Conference, Copenhagen, October.

Lundvall B.A. (1999) Technology policy in the learning economy, in Archibugi D. *et al.* (eds.) *Innovation Policy in a Global Economy*. Cambridge University Press, Cambridge.

Maillat D. (1995) Territorial dynamic, innovative milieus and regional policy. *Entrepreneurship and Regional Development*, **7**, 157–165.

Nelson R. and Winter S. (1982) *An Evolutionary Theory of Economic Change*. Harvard University Press, Cambridge.

Nelson R. (ed.) (1993) *National Innovation Systems: A Comparative Analysis*. Oxford University Press, New York.

Nonaka I. and Takeuchi H. (1995) *The Knowledge-Creating Company*. Oxford University Press, New York.

North D.C. (1990) *Institutions, International Change and Economic Performance*. Cambridge University Press, Cambridge.

OECD (1992) *Technology and the Economy: The Key Relationships*. Paris: OECD.

Ohmae K. (1995) *The End of the Nation State: The Rise of Regional Economies*. Free Press, New York.

Porter M.E. (1990) *The Competitive Advantage of Nations*. Free Press, New York.

Prusak L. and Cohen D. (2001) How to invest in social capital. *Harvard Business Review* **June**, 86–93.

Putnam R.D. (1993) *Making Democracy Work: Civic Traditions in Modern Italy*. Princeton University Press, Princeton.

Rallet A. and Torre A. (1998) On geography and technology: Proximity relations in localised innovation networks, in Steiner M. (ed.) *Clusters and Regional Specialization*. J H L DEWHURST.

Romano A., Elia V. and Passiante G. (eds.) (2001) *Creating Business Innovation Leadership. An Ongoing Experiment: The e-Business Management School at ISUFI*. Edizioni Scientifiche Italiane, Napoli.

Romano A., Passiante G. and Elia V. (2001) New sources of clustering in the digital economy. *Journal of Small Business and Enterprise Development* **8**(1), 19–27.

Sabel C. (1995) Experimental Regionalism and the Dilemmas of Regional Economic Policy in Europe, paper to international seminar on *Local Systems of Small Firms and Job Creation*. OECD, Paris, June.

Saviotti P.P. (2001) Networks, national innovation systems and self-organisation, in Fischer M.M. and Fröhlich J. (eds.) *Knowledge, Complexity and Innovation Systems*. Springer, Berlin, 21–45.

Saxenian A. (2000) Regional networks and innovation in Silicon Valley and Route 128, in Acs Z.J. (ed.) *Regional Innovation, Knowledge and Global Change*. Pinter, London, 123–138.

Scott A.J. (1998) *Regions and the World Economy: The Coming Shape of Global Production, Competition, and Political Order*. Oxford University Press, Oxford.

Seeman P. *et al.* (2000) Building intangible assets: A strategic framework for investing in intellectual capital, in Morey D., Maybury M. and Thuraisingham B. (eds.) *Knowledge Management: Classic and Contemporary Works*. The MIT Press, Cambridge, 85–98.

Senge P. (1990) *The Fifth Discipline: The Art and Practice of the Learning Organization*. Doubleday/Currency, New York.

Senge P. (2000) Reflection on a leader's new work: Building learning organizations, in Morey D., Maybury M. and Thuraisingham B. (eds.) *Knowledge Management: Classic and Contemporary Works*. The MIT Press, Cambridge, 53–60.

Storper M. and Scott A. (1995) The wealth of regions: Market forces and policy imperatives in local and global context. *Futures* **27**, 505–526.

Shapiro C. and Varian H.R. (1999) *Information Rules*. Harvard Business School Press, Boston.

Vonortas N.S. and Tolnay A. (2001) Towards the knowledge-based economy, in Feldman M.P. and Link A.N. (eds.) *Innovation Policy in the Knowledge-based Economy*. Kluwer Academic Publishers, Dordretch, 375–408.

3

A Knowledge Hub for Regional Development

Mario Marinazzo

3.1. Introduction

From the conclusions made in Chapter 1 regarding Virtual Industrial Clusters (VIC's), we highlighted the special need for:

- Regional production systems to be recognised increasingly, in the process of drawing up policies for supporting development, as economic units of fundamental importance in the globalisation age.
- The processes by which the regional economy develops to be supported on the basis of the two new paradigms of digital connectivity and "virtuality".
- In particular, the attention paid to learning processes based on the relationships between economic players to be proportional to their importance.

From the conclusions of Chapter 2, we also understand the need for regional development policies to deal successfully with the relationships between ICTs, institutions and regional development, with particular emphasis on two dimensions — network externality and social capital.

As regards the "Region Enterprise" metaphor introduced in Chapter 2, Sec. 1.4.1, we propose shifting the analytical unit for innovation away from products and processes towards the region concept and the region's ability to be a leader compared with its competitors. We are therefore in a position to state that at the heart of the debate we should be placing the capacity of a region to train, recognise and nurture leaderships capable of leading the area in its processes of innovation and change. In other words, paraphrasing the definition which Senge proposes for leadership within an organisation (2000), we introduce Regional Innovation Leadership (RIL) as the ability "to initiate and to sustain significant change to work effectively with the forces that shape change ...". This definition echoes the definitions which Senge himself proposes for *knowledge* — "... capacity for effective action ..." — and for *learning* — "... the process whereby human beings build knowledge, enhance their capacity for effective action ...".

In turn, the references to knowledge and learning bring us back to the concept of intellectual capital, and to its three parts, in the following definition (Seemann *et al.*, 2000):

- *Human capital* is the knowledge, skills, and experiences possessed by individuals. It comprises both explicit conceptual knowledge as well as more tacit knowledge.
- *Structural capital* includes the explicit, rule-based knowledge embedded in the organisation's work processes and systems, or encoded in written policies, training documentation, or shared data bases of "best practices". It also includes intellectual property, recognised by patents and copyrights.
- *Social capital* is reflected in the ability of groups to collaborate and work together and is a function of trust. Effective networks of relationships characterised by high levels of trust are a valuable and often overlooked resource in the creation and use of knowledge.

On the basis of the considerations set out above, in this chapter we will be describing the multiplicity of relationships which link regional development processes to intellectual capital, by proposing:

(a) A model for setting up and organising leadership in regional innovation.
(b) A pilot initiative managing knowledge for regional development.

We are assuming that the term "region" here means the space within which continuous exchange of information and ideas can take place between public and private organisations, which all share the same institutional, social and cultural context. This space is usually synonymous with the NUTS III level adopted by the European Union, or with other more detailed scales, at the level of regional communities at a sub-provincial level. For this reason, the attributes "regional" and "local" are here regarded as synonyms.

This chapter is divided into four main parts:

- We examine the conceptual basis of business innovation and knowledge leadership. We focus our attention on the regional government authorities and on the regional entrepreneurial communities as "forces which shape change". This will enable us to translate the concepts and methods commonly used in firms into local government practice, for the benefit of the firms located within the region.
- We study a series of contributions which delineate the key role played by knowledge as a primary factor in innovation and competitiveness, and we formulate a schema of the cycle of leadership in regional innovation.
- We list the values offered to the various types of organisation involved in regional development by an initiative that could be defined as a "Knowledge Hub" (KH) for regional development. We describe the basis for a local partnership between public and private organisations to promote, build and run such an initiative.
- We conclude by detailing the main functional specifications that distinguish a Knowledge Hub for regional development, linking each one to innovation development processes.

3.2. Leadership in Regional Innovation

We will be focusing our attention on the regional government bodies and on the entrepreneurial communities, because of their role as "forces which shape change" and their tendency to react to "actions" and external pressures. Insofar as the performance of the regional production system is conditioned by these factors, we maintain that both innovation and leadership (in their local dimensions) call for close attention. It is from this point of view that we pick up on the metaphor used in Chapter 2 of the "Region Enterprise", defining Leadership in Regional Innovation as "the collective ability of a local community to create and sustain significant changes in order to work efficiently with the force which shape change ...".

It is interesting to examine the most characteristic features of the positions which Senge refers to as "learning organisations", but applying them this time to a regional community:

- The *regional learning community* is a community in which individuals and groups continuously expand their ability to achieve the results they desire; in which new and expansive schemes of thought are nourished; in which individuals continuously improve competencies of "learning together".
- The *regional learning community* does not learn through a series of isolated episodes, but learns as it works to achieve the results it desires. In other words, it learns in order to improve its ability to successfully carry out actions.

In this sense, we should expect the regional community to be able to recognise the "forces shaping change" both inside and outside the region, in order to "work successfully with them".

Assuming the importance of knowledge and learning as fundamental assets to enable the regional community to face the dynamics of development in the globalisation age raises the question as to how the regional community should undertake specific initiatives in order to improve its performance in acquiring, creating, using and distributing knowledge that may be useful for development.

Let us answer the question by first of all outlining a reference model: the cycle of Regional Innovation Leadership.

3.3. The Cycle of Regional Innovation Leadership (RIL)

The basic question, when discussing regional and knowledge development, would seem to be what has to be done to boost the local communities' dynamic awareness of their competitive positioning, so that a common policy for undertaking or updating new collective actions can be found. This basic question, asked in these terms, leads to other issues, such as:

- How to empower the capacity of public decision-makers to conceive and realise development programmes related to production "clusters"?
- How to reduce the bureaucracy involved in taking public decisions which are connected with development?

- How to recognise the leaderships which are emerging in either the public or private sectors?
- How to recognise the region's specific knowledge base, which is difficult to reproduce elsewhere, and which could be the object of investment?

Dealing with all of these questions opens up a vast scenario of research, focused on two main concepts:

- The concept of intellectual capital in its three component parts — Human, Social, Structural — (Seemann *et al.*, 2000) and in its "community" dimension.
- The concept of the Dynamic Innovation System (Lundvall and Borras, 1997) or learning region (Florida, 1995).

Both these concepts affect the final question, i.e. whether or not direct intervention can in fact significantly modify a regional community's development path.

A series of positive answers to this question suggest that the development path can be modified by:

- Regional innovation systems characterised by "economic, political and institutional relationships" capable of generating collective learning processes "leading to the rapid diffusion of knowledge and best practice" (Nauwelaers *et al.*, 1995).
- The explicit adoption of regional development strategy models based on production clusters (Porter 1998, 1999, 2000).
- Regional government structures oriented in a novel way towards the construction of consensus and participation (Landabaso, Oughton, and Morgan, 1999).
- A more strategic and interactive way of approaching the definition and application of development policies based on innovation (Nauwelaers, 2001).

So, all of the possible "modification factors" of the regional development pathway are linked with knowledge, which finds new ways of connecting the ideas emerging from the learning economies with those coming from development based on clusters. This is a whole new family of regional development policies: policies directly connected with "political projects" (Mariussen, 2001) which involve leaders and public and private organisations, aiming to consolidate forms of knowledge government, including investments in training, research and innovation-directed collaboration.

In fact, policies aimed at knowledge government are also effective at orienting the education and training system towards new forms of production and maintenance of human capital: it is a known fact that today competitive advantage originates from market-oriented collaboration, which leads to learning and innovation, rather than from the quality and quantity of formal levels of teaching. Today, in effect, explicit knowledge (which is at the heart of formal education) is increasingly "mobile", so human capital with high levels of formal education is available almost anywhere.

The innovation networks promoted by local and regional governments, between entrepreneurs, universities and research organisations, put together concrete "projects" in which various "knowledges" co-operate in building and exploiting a competitive advantage in the marketplace. These "projects" originate from communities of entrepreneurs, institutional investors, "assemblers" of development ideas based on demand rather than on supply, public and private leaders. These projects increasingly originate from a regional rather than a national level. In other words, these "projects" originate from the regional community's intellectual capital, in its threefold form of human, social and structural capital.

The formation and operational development of such "projects" also generates tacit knowledge through "learning-by-operating": a type of knowledge that is extremely difficult to spread, and which is capable of being the true competitive advantage of a regional community.

On the basis of the above concepts, intellectual capital is fed by (and in turn feeds):

- Interactive learning and innovation.
- Development based on clusters.
- The growth of regional government bodies in their attempts to gradually reduce self-referentiality and institutionalised thinking.
- Understanding and applying the knowledge government mechanisms based on the market and on collaboration in the context of the Digital Economy.

Let us now identify the innovation leadership of a community in its ability to :

- Learn by interacting.
- Promote co-operation and networking between firms within the community.

- Define and carry out innovation policies linked to the concrete problems of regional production.
- Grasp the innate opportunities for change in the new processes made available by the Digital Economy.

These abilities are connected with the management of intellectual capital in its three manifestations — human, social and structural (Seemann *et al.*, 2000), while appropriate processes and policies for creating and managing intellectual capital produce innovation and increased competitiveness in the overlapping space shared by businesses and the local community. This is the space where:

- The decisions for new project-based economies come to fruition (Mariussen, 2001).
- Economic communities establish themselves through "strong, responsive relationships between the economy and community" (Wolfe, 2000).
- They are selected on a project-by-project competitive basis, and organisational change processes occur (Eliasson, 2000).

Recall that:

- *Human capital* is the knowledge, skills, and experiences possessed by individuals.
- *Structural capital* includes the knowledge embedded in the organisation's work processes and systems, or encoded in written policies, training documentation, or shared data bases of "best practices".
- *Social capital* is reflected in the ability of groups to collaborate and work together and is a function of trust.

The reading of "region as enterprise" suggests that the true novelty must consist in creating and managing social capital: a capital made up of the *relationships* between people and groups, rather than of people and groups as such.

For us here, the question of social capital is no longer just a means of interpreting historical processes (cf. Putnam, Robert D. 1993) but seeks to deal with how to intervene positively to ensure that social capital, along with people and technology, is something that can be invested in and managed.

From this point of view, a good starting point for a regional knowledge development strategy is to provoke and/or provide support for the relationships between people and/or groups variously responsible for social and economic growth, each time identifying the most suitable forms

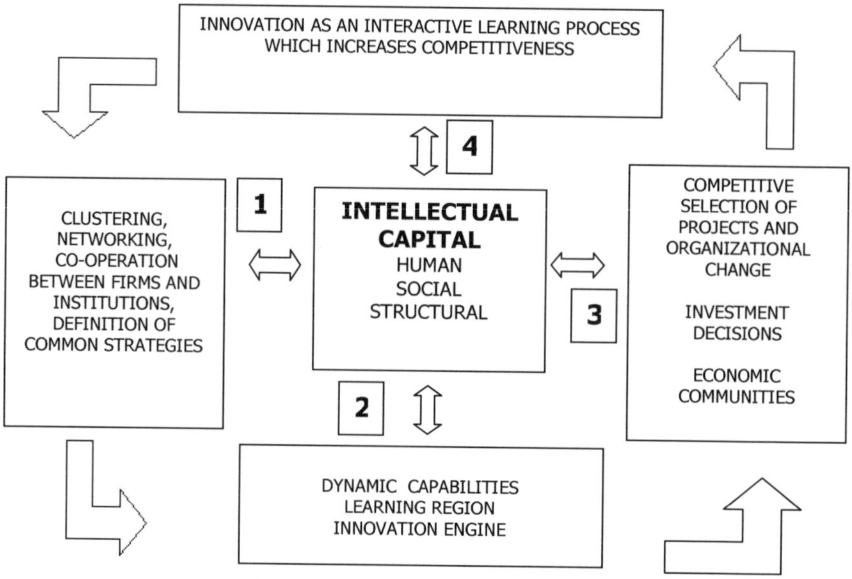

Fig. 3.1. The cycle of regional innovation leadership.

that will enable these relationships to be effective in terms of the objectives set out for regional development.

With the aim of using it as a reference for a knowledge-management pilot scheme, we can thus design a cycle of regional innovation leadership (Fig. 3.1) which summarises a variety of theoretical contributions which converge on the key role played by leadership in knowledge as a primary innovative and competitive factor.

The cycle starts off (and ends) with the competitiveness deriving from the innovation which has grown up through interaction in "action" contexts. Interaction is facilitated by collaborative processes among firms, and between firms and institutions (point 2 of the cycle).

In particular, the interactions which happen in "action" contexts and in problem-solving behaviours (point 3 of the cycle) increase dynamic capabilities (cf. Teece *et al.*, 1997) form the "learning region" (cf. Florida, 1995), and encourage local innovation engines to come forward (Hamel, 2001). In turn, these factors produce projects for change (cf. Eliasson, 2000), decisions for change (cf. Mariussen, 2001), and highlight economic communities which pilot the regional community in innovation processes (cf. Wolfe, 2000).

Intellectual capital is at one and the same time cause and effect of the processes which lead to innovation in the regional community, as, with reference to Fig. 3.1:

1. *Social capital* determines collaboration and regional clustering, as well as knowledge spill-over. *Structural capital* (such as a specific constructed or natural asset) determines the event at the origin of a regional cluster. *Human capital* (such as an "assembler" of development programs, i.e. a "catalyser" of consensus on a growth objective) is the "spark" at the origin of collaborations between entrepreneurs and institutions. Clustering reduces the production and management costs of collective infrastructures and collective assets in general (*structural capital*), but increases the efficiency of the regional specialist *human capital* market. Clustering reduces the costs of interacting for organisations in the cluster, thus favouring the growth of *social capital*.
2. *Social capital* feeds on "high performance routines", on which the *dynamic capabilities* of the regional community are based. These reside in the values, culture and organisational experience of the community. The *"learning region"* is characterised by:
   - a competitive base formed by continuous production of knowledge and improvements, stemming from the combination of the three components of *intellectual capital*;
   - a production system for goods and services based on the innate knowledge found in the components of *human and structural capital*, and whose value is optimized through the *social capital* component;
   - an industrial production infrastructure which seizes opportunities for innovation from the relational capital (i.e. the *social capital*) which connects regional producers among themselves and with their respective markets;
   - a system of continuous improvement of *human capital* through training initiatives contextualised for planning, production and marketing activities;
   - a system of infrastructures (*structural capital*) which can enable the acquisition, processing and production of constituent and reconstituent knowledge for *human capital*, and the relationships within the regional community and between the community and the global level (*social capital*).

The ability to learn builds up and maintains itself via processes of combining and recombining ideas and suggestions that arise from the interaction of individuals and groups who have different disciplinary competencies, values, experiences and knowledge: the three forms of intellectual capital are activated together in these processes.

The *innovation engines* are activated and operate via processes which "contaminate" heterogeneous human capital, and which stimulate and encourage creativity as well as technological, scientific and managerial competencies.

3. The social dimension (e.g. the mutual trust between public and private actors who commit themselves to a shared investment) and the structural dimension (e.g. the key elements of the regional innovation system) of intellectual capital favour the translation of this innate knowledge within human capital into economically-relevant processes. They thus lead to programs, projects, inter-institutional agreements, and between the public and private sector in support of these investment decisions.
4. Innovation processes are accompanied by processes in which new knowledge is produced and new professionality is created. Intellectual capital enriches itself with tacit knowledge, as well as the capacity to acquire and evaluate exogenous knowledge.

The cycle of Regional Innovation Leadership includes virtual clustering (Passiante, Elia, Massari, 2000) in its capacity as the main factor in attracting investments: "*... innovation and learning processes linked directly to economic and social growth are by nature interactive and collaborative, and involve individuals, groups and organisations. From this point of view, the "regional community-virtual cluster" metaphor goes beyond the level of regional marketing, and becomes involved in a wider field — that of economic and social development at a local level ...*" [ibid].

Our research hypothesis is that the RIL cycle can benefit from a "virtual" initiative based on IT and communications technologies:

- Oriented towards recognising and supporting "local champions".
- Involving networking facilitators.
- Aiming to promote and match up regional development strategies.

The term "virtual" ("... any work carried out over a distance of time or space, generally with the aid of electronic communication ..." (Cohen and Prusak, 2001)) derives from the desire to reduce the need for permanent organisational structures, and to make it easier to manage the net-based relationships aimed at achieving specific objectives.

Also, "... the global positioning of local economies requires policies and start-up mechanisms which reinforce the ability of regional systems to operate like the nodes in global networks. Global positioning can be achieved by selecting operational objectives and assigning resources to programs and actions which would validate the Virtual Innovation System at a regional level, through an improved ability to link at the local and global level, in order to broaden the spaces and opportunities for learning, thus increasing the potential for innovation ..." (Romano and Passiante, 1997).

Such a complex initiative has the particular features of a Knowledge Management System, as it is oriented towards "creating value from intangible assets" (Sveiby, 2001), such as human, social and structural capital. We call this knowledge management system as a "Knowledge Hub" (KH) capable of enabling innovation at a regional level by:

- Encouraging analyses and proposals in a problem-solving approach.
- Linking endogenous and exogenous knowledge sources.
- Supporting new dialogue, in a search for new prospects.
- Feeding discussions on the basis of competitiveness analyses and cluster/industry analyses.
- Encouraging innovative efficient practices, starting both from local and non-local experiences.
- Codifying the knowledge generated by analysing case studies.
- Linking endogenous and exogenous investment capacities.

In the next section, we will be outlining the values which can be obtained from the regional development-oriented Knowledge Hub.

3.4. An Outline of the Values Proposed for the Regional Development-Oriented Knowledge Hub (KH) Initiative

The various players involved in regional development have an interest in the process which stimulates innovation and expands into the regional production system, investment is attracted and competitiveness improves.

For *local authorities*, for *public decision-makers* and for *firms*, the value produced by the KH initiative originates in knowledge-sharing processes based on factual evidence, capable of reinforcing mutual trust through the shared consciousness and continuous dialogue which accompanies shared knowledge.

The sources of value for users and customers (local authorities and firms) include:

- The discussions and learning which arise from competitiveness analyses.
- The discussions and learning which arise from analyses of the structural and short-term characteristics at a sector and regional level.
- SWOT (Strengths, Weaknesses, Opportunities and Threats) analyses carried out with the help of interactive channels on the Web, such as:
  - forums;
  - on-line questionnaires;
  - knowledge diffusion operations using "push" technologies.
- Clustering among firms on projects of common interest as a result of knowledge sharing.

The value for the *University/Research* system originates from the new knowledge which is acquired by exploring how and under what conditions knowledge-sharing efficiently identifies and drives consolidated and/or emerging leaderships, and how well it does so in the varied context of the public and private tasks which are intrinsic to local development. The sources of value for the University/Research system include:

- The results of monitoring regional development.
- Experimentation "in the field" of methods for creating and extracting knowledge.
- The opportunity to package education, training and research activities close to the demand expressed by the regional community.
- The organisational modes suitable for transforming knowledge into competency: e.g. by boosting public decision-makers' ability to conceive and bring to fruition development projects which create value both for endogenous and exogenous assets.

The following section summarises the main features of the KH from the point of view of basic application functions.

3.5. The Regional Development-Oriented Knowledge Hub: Main Functions

The regional development-oriented Knowledge Hub aims to be an infrastructure which will enable the growth and transformation of intellectual capital in the regional community, i.e. a promoter of the ability to innovate.

These abilities can be summarised in the following points (Arthur D. Little Ltd., 2001):

- Vision and strategy.
- Competency-based management.
- Management of creativity and ideas.
- Culture and context.
- Intelligence.
- Organisation and processes.

The KH will propose analyses and diagnoses at a production sector and system level, in order to start up discussions and seminars capable of producing visions and strategies. If there is an understanding of the best practices used at a global level to resolve similar problems, this will be a factor in making it easier for visions and strategies to be forthcoming.

Regional core competencies will be measured and monitored and will also be benchmarked against other regions. Management of creativity and ideas will be promoted through exposing the viewpoints and proposals of public decision-makers and entrepreneurs for evaluation by the local community, supported by experts.

The KH will contribute to the dynamics of cultural change by constantly encouraging people to look at the problems of regional development in the light of the great transformations that are taking place, and to filter the possible solutions through experiences taken from a global context. *Intelligence* will enjoy growth opportunities by proposing analysis pathways guided by experts and consultants, on the basis of specific local problems, both using tools such as interactive forums, and with traditional forms such as workshops and meeting cycles.

The KH will contribute to the organisational ability by increasing the opportunities for entrepreneurs, public decision-makers, financial institutions and research teams to exchange views. These exchanges may lead to the emergence of connections (often neglected) between technology and finance, between innovative processes and internal competencies within organisations, between local and outside investors. These must then be adequately managed.

As knowledge is by its very nature a social fact, generated and shared through interactions in a problem-solving context, and these interactions take place in groups, the regional development-oriented KH will lever the potential of Communities of Practice (CPs), whose operativity should be promoted and sustained. The main characteristic of CPs activated through the KH has to do with being inter-sectoral, inter-institutional and inter-disciplinary.

The Regional Development-oriented Knowledge Hub must be grounded on principles of evidence and sharing of problems and related potential solutions. Problems should affect several social classes and business, including issues such as innovation policies and their impact, infrastructures, education, health and safety, access to external markets, etc.

CPs can be sustained by seven main phases in their problem-solving activities:

(a) Awareness.
(b) Animation.
(c) Codified knowledge sharing and diffusion.
(d) Feedbacks from knowledge interiorisation.
(e) Champions' recognition.
(f) Evidence of potential solutions.
(g) Implementation and monitoring.

The above mentioned seven-phased process focuses on issues such as the driving factors of:

- Regional competitiveness.
- Emerging problems and opportunities in specific sectors/clusters.
- Innovation infrastructure performance.

Regional Development-oriented CPs will be supported in adopting several problem solving methods, such as:

- Dialogue, seminars and meetings in both traditional and "virtual" ways (forums, newsgroups, etc.).
- Experience exchange with people from other Public Administrations, companies, business associations, universities, business services, etc.
- Continuing Knowledge gathering on relevant Regional problems, under the responsibility of recognised "gatekeepers".
- Inviting experts to participate in seminars.
- Diminishing barriers to knowledge acquisition taking small groups of regional practicioners to universities or elsewhere facilities and competencies are available to search in libraries, on the Web, etc.
- Appointing Communication professionals to help in presenting the results of a search session, a forum, a seminar, etc.

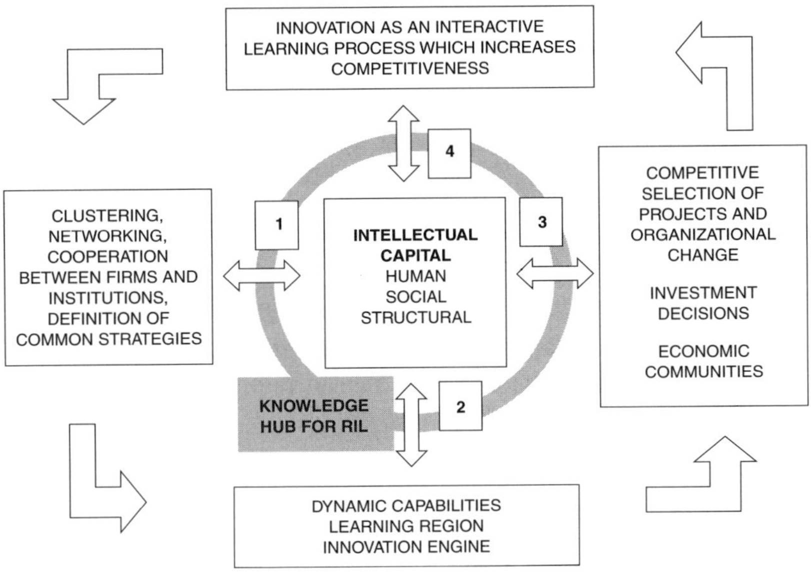

Fig. 3.2. The knowledge hub supporting the cycle of regional innovation leadership.

The Knowledge Hub, in order to sustain the cycle of leadership in regional innovation as defined above, will aim to sustain, promote and point out, rather than manage, the processes linking intellectual capital with the RIL cycle (see Fig. 3.2).

The functions of the KH will be focused on seven areas:

1. Knowledge sharing.
2. Regional innovation system.
3. Regional innovation systems at a global level.
4. Regional marketing.
5. Competitiveness of local systems.
6. Sector/Cluster analysis.
7. Regional development policies.

The regional development-oriented Knowledge Hub integrates functions for supporting the growth of intellectual capital with functions for transforming that intellectual capital.

With this in mind, below we directly compare the functions of Knowledge Management and the seven areas listed above, in a problem-solving logic, with the processes which link intellectual capital with the various phases of the RIL cycle.

Forums, discussion groups and on-line questionnaires (Knowledge Sharing area) will sustain the dialogue on problems and opportunities focusing mainly on:

- As regards the relationship between intellectual capital and clustering (see 1 in Fig. 3.2):
  - Prospects for collaboration and clustering.
  - Hypotheses for knowledge spill-over.
  - Hypotheses for setting up and developing collective assets.
- As regards the relationships between intellectual capital and the development of the abilities to learn, to distribute intelligence and to boost innovation leadership (see 2 in Fig. 3.2):
  - Highlighting the need for training within the context of activities for producing goods and services.
  - Opportunities to meet up with partners and suppliers at a regional and international level to discuss innovation hypotheses.
  - Opportunities to boost and recognise leaders, especially at a regional level, in development-related problem-solving processes.
  - Building of consensus and conditions of mutual trust and transparent collaboration, even between rival firms, as well as between heterogeneous companies, in order to achieve the objective of providing a common solution to problems recognised by all partners.
- As regards the relationship between intellectual capital and decision making (see 3 in Fig. 3.2):
  - Translation into projects and programs of the ideas which came to fruition from dialogue and social discussion.
  - Formal construction of partnerships aiming to find operational solutions to emerging problems and opportunities.
- As regards the relationship between intellectual capital and innovation (see 4 in Fig. 3.2):
  - Ongoing monitoring of innovation practices, in order to support collective learning processes and sustain feedback on projects and programs.

The part played by case studies and reports on the most efficient practices emerging in the area (Regional Innovation System area) will sustain the processes of awareness and collective learning, focused mainly on:

- As regards the relationship between intellectual capital and clustering (see 1 in Fig. 3.2)
  - Collaborations via networks.
  - Knowledge exchange between heterogeneous production sectors.
  - Competitive advantages provided by collective infrastructures.
- As regards the relationships between intellectual capital and the development of the abilities to learn, to distribute intelligence and to boost innovation leadership (see 2 in Fig. 3.2):
  - Impact of training initiatives in the framework of innovation programs.
  - Competitive advantage procured through collaborations among local firms, or even between local and outside firms.
  - Contributions from individuals and groups in driving innovation processes.
- As regards the relationship between intellectual capital and decision making (see 3 in Fig. 3.2):
  - Projects and programs of particular relevance for the development of innovation, and partnerships set up in order to carry out such projects and programs.
- As regards the relationship between intellectual capital and innovation (see 4 in Fig. 3.2):
  - Results from ongoing monitoring of innovation practices.

Access to information and documentation on production sectors and clusters at a global level, deriving both from practical experience and from theoretical contributions, favouring the experiences of sector and virtual clusters (Regional Innovation Systems at a Global Level), will sustain the processes of awareness and learning targeted towards the emulation and "customisation" of exogenous experiences. Information and documentation will focus on the same issues and items as in the Regional Innovation System area.

The diffusion of specialist information targeted at potential outside investors, and of reports on the opportunities which originate from the potential outside investors themselves (Regional Marketing area), will sustain the investment processes coming directly from outside, focusing mainly on:

- As regards the relationship between intellectual capital and clustering (see 1 in Fig. 3.2)

  - The regional clusters of greatest interest for potential investors.
  - The potential advantages for the region of investment programs proposed by players outside the region, sometimes in tandem with local firms.

- As regards the relationships between intellectual capital and the development of the abilities to learn, to distribute intelligence and to boost innovation leadership (see 2 in Fig. 3.2):
  - "Distinctive" knowledge incorporated in local production processes, of particular interest for potential investors.
  - Education and Training initiatives contextualised for investment programs proposed by outside investors.
  - Competitive advantage deriving from inward investments.
  - Information on the main economic players in the region, on their production and their supply networks.
- As regards the relationship between intellectual capital and decision making (see 3 in Fig. 3.2):
  - Inward investment projects and programs of particular relevance for the development of innovation, and partnerships set up in order to carry out such projects and programs.
- As regards the relationship between intellectual capital and innovation (see 4 in Fig. 3.2):
  - Results from ongoing monitoring of inward investment programs.

The diffusion of data and information deriving from statistical analyses and from studies and reports of a socio-economic nature, and regional competitiveness analyses (Local Systems Competitiveness) will feed the local dialogue processes, focusing mainly on:

- As regards the relationships between intellectual capital and the development of the abilities to learn, to distribute intelligence and to boost innovation leadership (see 2 in Fig. 3.2):
  - Transformations to a sector's knowledge bases (Foray, 2000) in terms of:
    - √ Coded and tacit knowledge.
    - √ Balance between competitiveness and non-competitiveness of the context.
    - √ Innovation rate, intensity of obsolescence and knowledge processes.

- As regards the relationship between intellectual capital and innovation (see 4 in Fig. 3.2):
  - The innovative performance of the local system, according to the criteria and methods of the European Innovation Scoreboard Initiative (EC-DG Enterprise, 2000) based on the following four dimensions:
    - √ Human resources;
    - √ Creation of new knowledge;
    - √ Transmission and application of knowledge;
    - √ Finance for innovation, outputs and markets.

The periodic diffusion of data and information on production sectors of particular importance for the local area will sustain the processes of awareness and learning (Sector/Cluster analysis area), focusing mainly on:

- As regards the relationship between intellectual capital and clustering (see 1 in Fig. 3.2):
  - The most significant regional production systems, highlighting the dynamics of competitiveness.
- As regards the relationships between intellectual capital and the development of the abilities to learn, to distribute intelligence and to boost innovation leadership (see 2 in Fig. 3.2):
  - Evolution of "distinctive" knowledge incorporated in local production processes.
  - Education and Training initiatives contextualised for increasing the competitiveness of a production sector.
  - Competitive advantage for the local community deriving from innovative investment in a particular sector.
  - Information, for each sector, on the main economic players in the region, on their production, on their markets and on their supply networks.
- As regards the relationship between intellectual capital and decision making (see 3 in Fig. 3.2):
  - Projects and programs of particular relevance for the development of innovation in a particular sector, and partnerships set up in order to carry out such projects and programs.
- As regards the relationship between intellectual capital and innovation (see 4 in Fig. 3.2):

- Results from monitoring development programs for a particular sector.
- Dynamics of the competitiveness of a particular local production system.

The periodic diffusion of data and information on policies for sustaining regional development (Local development policy area) will sustain the processes of awareness and learning, focusing mainly on:

- As regards the relationship between intellectual capital and clustering (see 1 in Fig. 3.2):
  - The policies explicitly aiming to promote and support clusters, and their impact.
- As regards the relationships between intellectual capital and the development of the abilities to learn, to distribute intelligence and to boost innovation leadership (see 2 in Fig. 3.2):
  - Education and Training programmes, contextualised for the development of regional production processes.
  - Policies based on co-financing investment in innovation.
  - Information on development policies promoted or managed at a regional, national and international level.
- As regards the relationship between intellectual capital and decision making (see 3 in Fig. 3.2):
  - Highlighting the opportunities for co-financing arising from public policies for sustaining innovations as well as from projects and programs set up at local level.
- As regards the relationship between intellectual capital and innovation (see 4 in Fig. 3.2):
  - Efficient practices of public support for regional competitiveness, especially with regard to studies and evaluation reports on behalf of institutions at a regional, national and international level.

The KH, supported by a partnership of local authorities, entrepreneurial associations, research institutions, education and training bodies and banks, will point its users in the direction of the services provided by the KH partners (one-stop shops for companies; information on opportunities for public and private co-financing of investment in companies; financial services, etc.).

Activities for promoting and diffusing the awareness of the opportunities provided by the Knowledge Hub will also be carried out, and these will be in the form of conferences, seminars, and consensus-building initiatives. All of these activities will be organised and carried out in conjunction with the partners in the KH.

3.6. Conclusions

This chapter has laid down a conceptual basis for an initiative setting up a regional-development oriented Knowledge Hub, and has identified the major functional characteristics involved. A prototype is currently being constructed, and experiments will begin very soon.

Its main aim is to promote the enrichment and transformation of the local development-oriented intellectual capital within the region.

In order to achieve these aims, the initiative is directed above all at the operational processes and at the documented and documentable facts which accompany the production and marketing of goods and services, management of the economy and of the area, and the supply of basic services (first and foremost Education and Training).

By focusing on operational processes, problems can be linked to proposals, through dialogue and social discussion between "competent" people and groups.

Focusing on operational processes also makes it possible to:

- Select the explicit knowledge needed to sustain dialogue and argument about the problems which are identified.
- Activate the necessary processes of coding knowledge.
- Contribute to building knowledge as ability to act efficiently in favour of regional development.

There are several conditions affecting the outcome of the initiative, and the most important of these are not technological: in fact, ICTs, as enablers of the Digital Economy, can also carry out the typical functions of a KH. The social and cultural conditions are overwhelming. So, the institutions must play a role that is vast, profound and convinced.

The metaphor of "region as enterprise" has led us to draw up this proposal, despite the incompatibility of some characteristics of a firm compared with those of a region, as mentioned in Chapter 2. Above all, we are eager to underline that traditional innovation policy seems now to be stabilising in practically automatic forms of incentives for innovation, concentrating on material aspects (infrastructure, research, technology, finance) and on immaterial aspects such as education and training.

The lesson which comes from the literature on corporate competitiveness increasingly highlights knowledge as the ability to act effectively, and learning as the readiness to acquire this ability. So, the regional development oriented KH offers itself as an innovation policy tool which is purely organisational, focusing on strategies, networks, culture, ethics and institutions: factors themselves capable of speeding up the effects of traditional investment. In other words, the regional development oriented KH aims to be a tool for change in those contexts where deficiencies in terms of networks, culture, ethics, institutions and strategies seem likely to thwart more traditional investment.

References

Arthur D. Little Ltd. (2001) *The Innovative Company: Using Policy to Promote the Development of Capacities for Innovation — Final Report to Participating Governments*. February.

Cohen D. and Prusak L. (2001) *In Good Company: How Social Capital Makes Organisations Work*. Harvard Business School Press.

Eliasson G. (2000) The role of knowledge in economic growth, paper presented to HRDC, Canada-OECD symposium on *The Contribution of Investment in Human and Social Capital to Sustained Economic Growth and Well-Being*. Quebec City, Canada OECD, March 19–21.

European Commission, DG Enterprise, Innovation and SMEs Programme (2000) *European Trend Chart on Innovation — The European Innovation Scoreboard — Commented Indicators*. EC, DG Enterprise.

Florida R. (1995) Toward the Learning Region. *Futures*, **27**(5), June, 527–536.

Foray D. (2000) Characterising the knowledge base: Available and missing indicators, in *Knowledge Management in the Learning Society*. OECD.

Hamel G. (2001) Innovation: The new route to new wealth, The Peter F. Drucker Foundation for Nonprofit Management. *Leader to Leader*, **19**, Winter.

Landabaso M., Oughton C. and Morgan K. (1999) Learning regions in Europe: Theory, policy and practice through the RIS experience, presented at the 3rd Annual Conference on Technology and Innovation Policy, Austin, Texas, September.

Lundvall, B.-Å. and Borras, S. (1997) *The Globalising Learning Economy: Implications for Innovation Policy*. DG XII-TSER, Bruxelles.

Mariussen, Age (ed.) (2001) *Cluster Policies: Cluster Development?* Nordregio Report:2, Stockholm.

Nauwelaers C. and Reid A. (1995) *Innovative Regions? A Comparative Review of Methods for Evaluating Regional Innovation Potential*. Commission des Communautés Européennes and RIDER, Louvain-la-Neuve, pp. 160.

Nauwelaers C. (2001) Path-dependency and the role of institutions in cluster policy generation, in Mariussen A (ed.) *Cluster Policies — Cluster Development?* Nordregio Report:2, Stockholm.

Passiante G., Elia V. and Massari T. (2000) *Net Economy Cacucci*, Bari.

Porter M.E. (1998) Clusters and competition: New agendas for companies, governments and institutions, in M.E. Porter (ed.) *On Competition?* Harvard Business Review Books.

Porter M.E. (1999) Clusters and the new economics of competition, *Harvard Business Review*. November–December, 77–90.

Porter M.E. (2000) Location, competition and economic development: Local clusters in a global economy. *Economic Development Quarterly* **14**, 1, February, 15–34.

Putnam R.D. (1993) *Making Democracy Work: Civic Traditions in Modern Italy*. Princeton University Press.

Putnam R.D. (1993) The prosperous community: Social capital and public life. *The American Prospect*. **13**.

Romano A. *et al.* (1997) Un modello per la gestione innovativa dei sistemi economici locali: il sistema innovativo virtuale, VII Convegno internazionale di economia e politica dell'innovazione, su Politiche pubbliche a sostegno dell'innovazione, Universitá Cattolica del Sacro Cuore, Cremona, 11–13 Giugno.

Seemann P. *et al.* (2000) Building intangible assets: A strategic framework for investing in intellectual capital, in Morey D., Maybury M. and Thuraisingham B. (eds.) *Knowledge Management*. The MIT Press, Cambridge Massachusetts.

Senge P. (2000) Classic work: The leader's New Work: Building learning organisations, in Morey D., Maybury M. and Thuraisingham B. (eds.) *Knowledge Management*. The MIT Press, Cambridge Massachusetts.

Sveiby K.E. (2001) www.sveiby.com.au/KnowledgeManagement.html.

Teece D.J., Pisano G. and Shuen A. (1997) Dynamic capabilities and strategic management, *Strategic Management Journal*, **20**, 509–533.

Wolfe D.A. (2000) Social capital and cluster development in learning regions, XVIII World Congress of the International Political Science Association. Quebec City, Quebec.

4

Net Economy Business Models for SMEs Operating in Fragmented Markets

Tommaso Massari

4.1 Introduction

This chapter of the book is devoted to exploring new e-Business strategies and models for SMEs operating in highly fragmented markets. The aim is to provide readers with a general overview of how the Net Economy impacts on strategies, business and technological models of industrial clusters operating in traditional sectors. The chapter emphasises the role of "federated distributed information systems" and "peer-to-peer" (P2P) business models and technologies in forging and shaping the new business arena for SMEs operating in these sectors.

The first part of the chapter describes new trends in e-Business model design based on a more distributed architecture of data, services and systems. Based on some preliminary assumptions, the work attempts to analyse potential e-Business scenarios for SMEs operating in traditional sectors. This part of the chapter describes an e-Business model —"OpenMarkets" — developed by the eBusiness Management School in conjunction with other partners in response to an EC initiative and based on peer-to-peer architectures, to accelerate the impact of Net Economy-led transformations on European SMEs.

The second part of the chapter describes a distributed system of peer-to-peer web services applied to the Tourism sector — regarded as "information intensive" and highly fragmented.

The chapter closes with some observations on new competitive strategies for SME clusters operating in fragmented markets such as tourism and suggests a way to speed up SME transition towards the Net Economy.

4.2 The "OpenMarkets" Model For European SMEs

4.2.1 *A new technological and business scenario for SMEs*

The Internet was first recognised as a way to market products to consumers. Companies like Amazon.com and Yahoo are two prime examples of B2C e-Commerce. Early examples of business-to-business interactions were based on the very same catalogue-driven functions. Currently all B2B solutions providers are focused on catalogues and exchange functionality whereby somebody provides the platform and technology for business to interact in a controlled environment. Soon it became apparent that the biggest gains would not come from lower prices or better liquidity but from integration along the value chain. Until now this has become reality only for big players. SMEs, especially, lack the influence to model the information flow after their workflows, and they are therefore forced to adopt workflows from their bigger competitors, which are normally ill suited for them.

In parallel, P2P applications have started to revolutionise the exchange of information and computing power. Napster, Gnutella, Freenet and SETI@home have been recognised by the media and investors alike, despite the fact that Napster was a hybrid and SETI@home peers do not interact with each other but only with a co-ordinating computer. Research institutes are expecting P2P-based applications to revolutionise the B2B-markets. However, at this point there are no existing technologies to support business interactions on P2P platforms.

As a logical expansion of existing component frameworks, agent platforms allow agents to provide and use services. They standardise the communication between these agents on a high level of abstraction. Existing agent platforms have been developed in research initiatives. The Foundation for Intelligent Physical Agents (FIPA) has proposed an agent standard reference framework. Despite the fact that a number of implementations of this standard exist, they have not found widespread use outside of research and development projects.

In this section we will describe a new business and technological model — "OpenMarkets" — that will combine the simplicity and flexibility of P2P applications with the power of existing B2B-applications.

The development of this model has been based on some assumptions regarding future technological and business scenarios:

Existing centralised electronic exchanges are ill suited for the needs of small and medium enterprises

The Internet is characterised by the Http-protocol and its associated web-servers. Since electronic marketplaces have developed as a special kind of website, all existing systems are based on this centralised technology. These exchanges provide services such as partner finding, pricing, and settlement. A number of technology providers, such as Ariba, CommerceOne, and Intershop provide standard solutions for these kinds of centralised exchanges.

As a consequence of the centralised technology and the necessary upfront investments, three classes of business models are most common amongst these exchanges:

- exchanges by a consortium of dominant buyers looking for lower cost of goods and increased efficiency in their procurement processes;
- transaction fee-based exchanges with investments from independent investors;
- exchanges sponsored by service providers as a value-added service to their customers.

Exchanges led by a consortium of big participants are very successful, if they operate in vertical markets that are dominated by a small number of buyers or sellers, e.g. Covisint for the automotive industry. For many other markets, the exchange business model has not been successful.

P2P — the new paradigm in Internet computing

Personal computers have penetrated society over recent years. Not only has the number of these computers been rising dramatically but also their power, memory and storage capacity. In parallel the development of the Internet has created high-bandwidth networks that connect all these CPUs. As a result, computing power has moved away from centralised servers to clients. However, until now, most of this computing power is unutilised.

P2P file sharing services have shown a potential use of these resources. The successors of Napster, initiatives like Gnutella, have offered the same file sharing functionality without the Achilles heel of the centralised catalogue. Consequently, they are more difficult to control but also do not provide the same quality of services. New approaches to this problem have not yet been developed.

A second direction for the application of P2P technology has been supercomputing. Distributed computing, as it is called, holds the promise of nearly unlimited computing power. As with all parallel computing the problem is not raw computing power, but the intelligent distribution of data and the modelling of the problem space which have to happen prior to the distribution.

As of late, companies like Groove have started to use peer-to-peer technologies to share information between different individuals. These applications are a direct extension of the services currently offered by centralised servers, e.g. MS Exchange Server. Therefore, they focus on small groups of individuals sharing information amongst each other.

The potential of P2P technology for B2B-application is widely recognised. But to date none of them has delivered a product or even shown a working prototype. A number of innovations are necessary to make P2P infrastructures suitable for B2B-applications. A later paragraph will discuss these innovations in more detail.

Realising the high potential of P2P, major companies have announced P2P initiatives without disclosing specifics about their plans. Sun Microsystems will publish their plans for P2P shortly. Microsoft has announced that it is to add a P2P component to its .NET initiative but not disclosed how the two concepts will relate.

Agent technologies

Agents are an approach to providing services in distributed networks. Agents represent autonomous software entities, which are characterised by pro-active and intelligent behaviour. Agent platforms are software frameworks which allow the deployment of agents and facilitate the standardised communication between agents on different peers. A reference architecture for such agent platforms has been specified by the "Foundation for Intelligent Physical Agents" (FIPA). FIPA compliant platforms provide the necessary infrastructure for agent-based interactions. Most importantly, they support communication between different agents and monitoring of agents via an "Agent Management

System". A directory facilitator provides yellow pages functionality for discovering available agents and their services.

For the communication between different agents, FIPA defined an extensible set of standard protocols and the agent communication language ACL. Protocols are defined as an expected sequence of speech acts, the basic elements of a dialogue. The syntactic elements of these speech acts are expressed in ACL.

A number of publicly available implementations of the FIPA standard exist, most importantly Grasshopper, FIPA-OS, JADE, and ZEUS. Unfortunately, these agent platform implementations do not show the necessary performance for distributed P2P applications; in addition, they are heavy-weight and difficult to administrate. Hence, they cannot easily be applied for P2P-based B2B applications. Agent-based marketplaces have been developed for research and commercially. However, they are bound to a single platform and therefore lack the flexibility, scalability, and cost benefits which a distributed marketplace offers.

4.2.2 *e-Business in fragmented markets*

The rapid development of the Internet, increasing globalisation and the development of service economies are creating the opportunity and necessity to integrate business partners along the value chain. The biggest gains are expected from c-Commerce, the collaborative integration of partners through the technology of the Internet. At present, exchanges and marketplaces represent the dominant approach to this kind of value chain integration. The existing centralised exchanges are shaped by and cater to established large corporations. These big players either dominate the buy or sell side of the market and thereby the exchange. Alternatively they act as sponsors of the exchange cross selling their services to market participants. The necessary upfront investments for these systems are prohibitive for the development of such marketplaces by and for small and medium size enterprises (SMEs). Therefore, SMEs have not been able to reap the benefits of c-Commerce and its associated technology.

The OpenMarkets model we propose carries a collaborative electronic marketplace characterised by the free exchange of information and transactions between equal peers. We propose a P2P network as the foundation of the OpenMarkets infrastructure. This infrastructure will offer all market participants including small and medium enterprises

- the ability to offer, buy and sell products;
- the flexibility to incorporate services required around the sales and procurement processes;
- an infrastructure requiring minimal upfront investments by leveraging existing Internet initiatives and open source technology;
- increased benefits from local networks and communities.

OpenMarkets is based on technologies combining existing P2P platforms with modern agent platforms. It supplies components, which provide P2P functionality, security mechanisms and basic transactions. OpenMarkets extends existing protocols towards B2B-applications and transactions. Thereby, agents will be able to provide application specific user interfaces, new transaction types, or automated negotiations. Specialised agents will enable principals to utilise the services of independent third parties, e.g. settlement or payment services.

Markets in a P2P world

In fragmented markets, buyers do not have access to all existing offerings. Furthermore, buyers might not be interested in all offerings — possibly due to logistics, legal or other circumstances. OpenMarkets will provide a consolidated view of the offerings and increase market transparency. Each market participant can restrict his selection by customising the consolidating agent.

Transactions in a P2P environment

At the center of any B2B interaction is the ability to transact. Existing P2P platforms do not support transactional structures. In order to enable businesses to transact via a P2P platform, OpenMarkets will define a generalised transaction protocol suitable for a P2P platform. Furthermore, agents for fixed price offerings and auctioning will be developed as show-cases for other transaction types and negotiation strategies.

Services on a P2P platform

Big efficiency gains are expected to come from the integration of different business partners along the value chain. Independent third parties can provide services, e.g. settlement, and thereby allow enterprises to outsource non-core functions. Consequently, corporations can focus on their core business and increase the overall efficiency of their processes.

Any B2B-platform has to enable participants to offer services to others; thus, special protocols will have to be developed and implemented to allow market participants to utilise services from independent third parties. These services will be advertised via the use of a directory facilitator. For instance, a company-internal index of suppliers can be made available to other interested parties as a web based service — either for free or subject to charge — and can be accessed by third parties' own peer devices or computers.

In addition to already established service providers, the use of a computer based marketplace enables the offering of special services like translation between ontologies or natural languages, EAI or ASP integration. Such services will make it possible for the marketplace to grow across industry boundaries, integrating multiple formerly fragmented markets into a bigger community.

Local networks

Enterprises rely on established relationships with business partners. Exchange-based e-Business cannot naturally support these valuable "local" networks. The OpenMarkets P2P platform will support these local networks. Market participants will be able — but not limited — to request bids from a limited group of business partners. Thereby, these business partners can act as intermediaries to business partners of theirs by forwarding the request for bids. Organisations with close business ties will be able to leverage their long-standing partnership to new levels and into new business areas.

Security

Security and authorisation mechanisms are essential components of a B2B-platform for businesses to transact over the Internet. Consequently, the development of a security framework will be a critical component for OpenMarkets. Message encryption and authentication will be important functionalities provided by such a security framework.

On each peer authorisation, mechanisms will enable the platform provider to tightly control the information and services it is disclosing. The necessary mechanisms have to be integrated into the component-based communication platform on multiple levels.

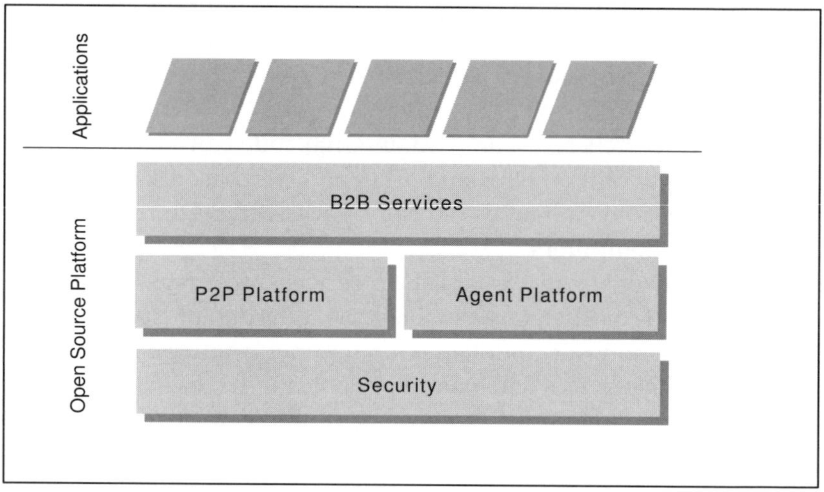

Fig. 4.1. Components of OpenMarkets.

4.2.3 OpenMarkets, an open-source B2B marketplace

The goal of the OpenMarkets model is the development of a B2B marketplace infrastructure based on open source P2P technology. Figure 4.1 shows the components of the platform and their relations to applications. These components are:

- **P2P layer** providing the necessary functionality for efficient collaboration based on decentralised services (e.g. network management, cascading search);
- **Agent environment** providing the means for the deployment and execution of software agents;
- **B2B layer** providing the specific services for OpenMarkets, the decentralised B2B market, e.g. transactions;
- **Security layer** providing mechanisms for encryption of messages, authorisation and authentication services.

The existing technology requires centralised structures for e-Business platforms with the discussed unfavourable consequences for SMEs. Merging the P2P paradigm with agent technology will make it possible to establish decentralised electronic markets. This approach promises great advantages in different regards.

OpenMarkets allows SMEs to shape and participate in B2B marketplaces

P2P infrastructures do not rely on centralised services and/or infrastructure. As such no market participant can dominate the technology. Every participant has the full flexibility to decide which products and services he presents, how and, most importantly, to whom he presents.

P2P markets can be extended by agent functionalities to support flexible, higher level services. An example shown in Fig. 4.2 is the pre-selection of offers through an agent, either a real person or a software agent, if a market participant delegates the information-gathering process. The resulting consolidated view onto the offerings by multiple sellers aid efficient decision making. If the settlement of the transaction is delegated to another agent, the integration along one step of the value chain is completed. Furthermore, it is possible to perform complex transactions across multiple levels of the value chain.

Therefore, agent-based P2P platforms are ideally suited for low cost, flexible systems for fragmented markets, in which most small and medium enterprises operate. For this reason, P2P-based B2B resembles the vision of c-Commerce as collaborative e-Business interactions within companies, between companies and their business partners, and among all parties throughout a trading community more than centralised exchanges.

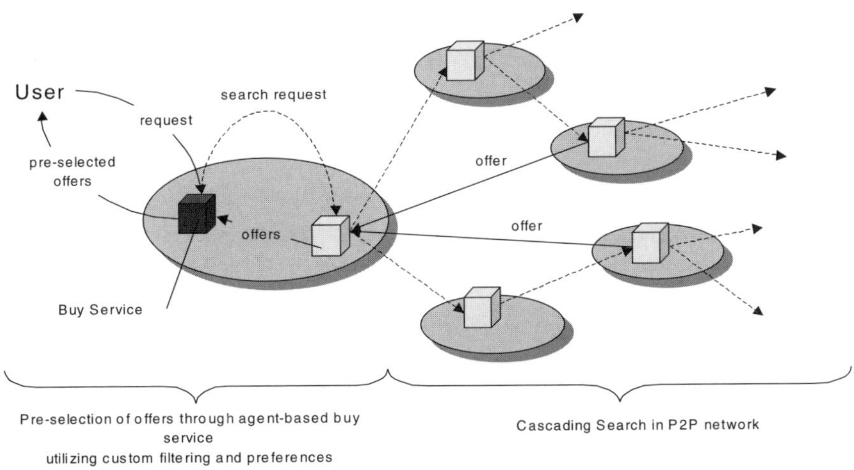

Fig. 4.2. Propagation of search request through a agent-based P2P network.

Integrating agent and P2P functionality

In our approach, we extend the agent platform by a P2P layer providing the P2P specific services. This layer will include components to manage P2P protocols and to perform distributed searches in the network. Additional services include mechanisms for access control and dynamic group configurations as well as persistence modules to allow for transparent storing of content and states. In the P2P scenario, intelligent agents extend the available mechanisms for communication in distributed environments by means of the seamless integration and use of services and the delegation of computing tasks to intelligent software entities. Hereby, additional transparency and efficiency is gained.

Content and ontologies

The exchange of information between peers necessitates consistent content descriptions. XML has emerged as the standard for the description of content and its syntax. However, the message content can be interpreted correctly only if an ontology is given to formally define the agreed and common knowledge in terms of concepts and relations. Different ontologies are necessary for different applications. Automated services for the translation from one ontology to another will have to be developed on top of the B2B infrastructure. There exist a number of initiatives for the definition of application specific ontologies, such as fpML (Financial Products Markup Language, www.fpml.org) and ebXML (Electronic Business XML from Oasis) [FEN00].

B2B services on a P2P platform

The integration of P2P technology and agent technology will allow peers to offer and utilise B2B services. The aforementioned translation services are one example of such a service. Other examples are settlement, clearing, information brokerage. Existing P2P platforms do not allow the integration of such generalised services. The ability to offer multiple and diverse services on the platform is crucial for the applicability to different application domains in B2B. Protocols and mechanisms for such service provision will be developed as part of the project.

Security

B2B applications require secure communication on all levels. Furthermore, for reliable transaction mechanisms, authentication is a pivotal requirement. To be accepted by businesses, the platform therefore has to supply sufficient security functionality on various levels: basic encryption, authentication and non-repudiation. The project will define the requirements for the obligatory security mechanisms. In a distributed environment, this goal can only be achieved by the usage of digital certificates in a "Public Key Infrastructure" (PKI). Regarding the development of the European law, the so called "qualified digital signature" has the same legal status as the traditional signature. The different European countries are already implementing suitable local regulations but the fact that no global PKI standards have been developed until now complicates the utilisation of PKIs for distributed networks.

Mobile access to an agent-based B2B market

The integration of mobile devices into a B2B marketplace can highly increase their availability and usefulness. Due to their low computing power and other limitations mobile devices will in the near future not be able to carry the full responsibility of a peer. On the other hand a pro-active agent can be monitored and controlled via a mobile device. Agents are therefore ideally suited to connect mobile devices to a B2B marketplace. The technology necessary for the delegation of roles and responsibilities to agents through a mobile device has to be developed. This includes the evaluation of existing protocols and technologies, and the implementation of an agent for such a mobile client.

4.2.4 *Benefits of the "OpenMarkets" model for European SMEs*

In several ways OpenMarkets technology contributes to the transition of European SMEs toward the Net Economy:

- OpenMarkets fosters the development of a novel internet technology based on P2P infrastructures which is dedicated to providing scalable and flexible networks.
- OpenMarkets creates an ambient intelligence landscape for seamless delivery of services and applications for B2B in Europe. OpenMarkets

redefines the relations between suppliers and consumers across the value network, leading to a global digital economy where consumers and businesses can seamlessly and dynamically come together. These value constellations can be assembled dynamically in response to constantly changing, highly customised market demands.

- OpenMarkets provides an open source infrastructure available for everyone and allowing any enterprise to participate in e-Business transactions and activities, the only cost being that of providing its own information in the P2P marketspace.
- With the incorporation of security models on the fundamental layer, OpenMarkets provides the necessary trust level for SMEs to manage business transactions over the Internet.
- OpenMarkets provides a basis for a number of innovative business models which can and will evolve based on the P2P infrastructure. SMEs can integrate their services seamlessly into the P2P network structure.
- OpenMarkets includes the vision of a standardised P2P infrastructure for different application fields. As such, it provides a bundling of the leading European forces in the emerging P2P/B2B area and will result in an open framework for fair competition and rapid innovation.
- OpenMarkets anticipates the needs in the B2B market which are currently evolving due to the failure of B2B exchanges and the corresponding business models. As such, the OpenMarket initiative builds a critical mass ahead of market developments.
- OpenMarkets will strengthen the competitiveness of SMEs in the targeted application areas in the worldwide competition in the P2P/B2B market. It will support the competitiveness of these SMEs by taking profit of more flexible e-Business and more effective B2B models based on the OpenMarkets technology. As such, public funding of this enabling technology will have a substantial impact.
- OpenMarkets will not only support SMEs in the targeted application areas, the OpenMarkets technology will give European SMEs in the IT sector in general — not only the participating SMEs — the chance to develop new business models based on the customisation, extension, and marketing of the Open-Source B2B software and to provide specialised services for SMEs in different application areas.
- OpenMarkets will allow SMEs to shape and use a marketspace independent of dominating market participants controlling the exchange. Existing B2B exchanges have to cater to big corporations

to create liquidity and transaction flow and thereby income from transaction fees. SMEs do not necessarily profit from the transparency of exchanges for standardised products.

- OpenMarkets will make application integration possible for SMEs through open source initiatives. The integration of processes is a key contributing factor to cost savings through e-Business. The ability to receive information on implementation details from the open source community facilitates the development of adaptors to other legacy systems. Possibly third parties will provide adaptors for standard software used by SMEs under an open source license.

Lowering barriers to successful e-Business initiatives for SMEs

The OpenMarkets model is in line with the EC GoDigital programme's overall purpose which is to support activities in order to help SMEs to use information and communication technologies (ICT) with the best possible efficiency. There are now over 19 million SMEs in Europe. In most EU Member States they make up over 99% of enterprises. SMEs, in general, will benefit from lower barriers to entry into markets as a consequence of e-Business technology. Hence, e-Business is often described as the SME's gateway to global business and markets. However in Europe there is a generally positive correlation between the size of an enterprise and its Internet use for business. As a result, the preponderance of SMEs in Europe's economy is not matched by their adaption of e-Business tools. Among all the constraints to the SME's digitalisation, the European Commission highlights the lack of technology and standards and the costs related to ICT investments.

The OpenMarkets model addresses these challenges by providing an open source P2P platform which could eventually lead to a standard B2B protocol. Providing a freely available entry solution for B2B initiatives of SMEs while still allowing for an upscale to the connection of legacy systems greatly lowers the ICT investment required by SMEs.

These benefits can be realised only if a critical mass of SMEs adopt such a standard to interact with each other and therefore a European scale project is needed to reinforce this process. By adopting open source platforms, SMEs will have a means to describe their services and business processes in a global, open environment on the Internet, thus extending their reach. Potential trading partners will quickly and dynamically discover and interact with each other via the Internet and their preferred applications, thus reducing time to market. The barriers to rapid participation in the global Internet economy will be lowered for any business anywhere, thus allowing them to fully participate in the new Digital Economy.

The first prototype application for OpenMarkets will consequently focus on two markets typically shaped by SMEs, agrifood and tourism. These markets are also most important in European regions with low levels of industrialisation.

Exploitation of new business opportunities

It is widely accepted that collaborative distributed environments on a P2P basis will play a very important role in future e-Business and B2B applications. OpenMarkets picks up on the P2P paradigm at a very early point in the development of this technology and includes very new approaches to enhance P2P technologies to provide value-added services for B2B applications. Based on this technology, OpenMarkets opens up new opportunities and allows for new business models in the B2B arena. Different from today's exchanges, various kinds of specialised services offered on OpenMarkets can easily be integrated into the business processes and the value chain. Examples of such services are application domain specific catalogues and translation services. In particular, these services can be provided by flexible young companies with a good understanding of the application domain, occupying emerging niches. Since the possible range of services for P2P-based B2B marketspaces is wide, the opportunities for the European industry are immense. This is especially true since the potential of value-added services for P2P networks has not yet been identified in the United States and Japan.

Besides the opportunities for service providers in the value chain, OpenMarkets also opens up new markets and allows for new business models in the IT sector. The open source OpenMarkets platform will provide the entry level peer for SMEs to participate in distributed marketspaces. B2B initiatives based on OpenMarkets can generate efficiency gains by connecting to the ERP systems. Similarly, company services need to be transposed and implemented to be made available in the marketspace. Both tasks can only be performed by software companies and consultants with a good understanding of the OpenMarkets platform. Consequently, OpenMarkets offers new business opportunities in the IT sector. Similarly to the effects of other open source initiatives such as Linux and Apache and recently Java, the consortium expects OpenMarkets to be an enabling technology for new business models and services.

Standardisation

OpenMarkets promotes the vision of a standardised P2P infrastructure for different application fields. As such, it provides a bundling of the leading European forces in the emerging P2P and B2B arena and will result in an open framework for fair competition and rapid innovation.

OpenMarkets will not only support SMEs in the targeted application areas. The OpenMarkets technology will give European SMEs in the IT sector in general — not only the participating SMEs — the chance to develop new business models based on the customisation, extension, and marketing of the Open-Source B2B software and to provide specialised services for SMEs in different application areas.

Increasing SMEs competitiveness

SMEs generate a substantial share of GDP and are a key source of new jobs as well as a breeding ground for entrepreneurship and new business ideas. Table 4.1 shows that despite the fact that many SMEs are having access to the Internet and are using it for distribution and information, the percentage of SMEs which generate value-added from the Internet through integration and cost savings is significantly lower than among larger corporations. Providing SMEs with open market platforms

Table 4.1. Use of the Internet by size of enterprise, Europe-1999.

Number of employees	1	1–9	10–49	50–49	Total
Having direct access to the internet	33%	49%	67%	86%	42%
Distribution of products information via the Internet	14%	27%	42%	59%	21%
Distribution of products via the Internet	6%	7%	9%	13%	7%
Receiving orders	8%	10%	15%	20%	10%
Confirming orders	6%	9%	12%	16%	8%
Paying via the Internet	4%	5%	8%	9%	4%
Receiving payment via the Internet	3%	3%	4%	7%	3%
Co-operating in offering products or services jointly	9%	13%	19%	29%	12%

Source: ENSR Enterprise Survey 1999 as used in The European Observatory for SMEs — Sixth Report, European Commission, 2000.

will unleash their potential by accelerating their business transformation process towards the e-Business which will improve their economic productivity and organisational efficiency. SMEs which are currently striving to survive in a global, more complex business environment will find new ways to co-operate and compete within a dynamic context. This will be reflected in a more distributed dynamic equilibrium between SMEs and large enterprises thus allowing for a better distribution of business knowledge and eventually create new job opportunities.

An open B2B architecture can contribute to the creation and reinforcement of co-operation networks between firms or groups of firms, research centres and universities, organisations responsible for improving the quality of human resources, financial institutions, and the dissemination of research results and technological adaptation in SMEs. This will enable them to reduce their disadvantages (in particular peripheral location) and leverage their advantages. Consequently, through its powerful networking effects the information society can contribute to the geographical and economic opening-up of these regions. The project will raise the technological awareness and market integration in micro-companies in the crafts and other traditional sectors.

4.3 A Federated-Distributed Model for the Tourism Sector

4.3.1 *The nature of the travel industry*

Global travel and tourism — one of the world's largest "industries" — generates $4.7 trillion in revenue annually, employs 207 million people either directly and indirectly, and supports hundreds of thousands of businesses.[1] The substantial growth of the tourism activity marks this sector as one of the most remarkable economic and social phenomena of the past century, not only in Europe but all over the world.

However, the impact of the internet revolution has profoundly changed the buying side of the tourism and travel value chain, in other terms the way consumers interact with the selling chain, but this shift towards digital channels has hardly affected the way tourism and travel products and services are traded in the supply chain (B2B).

---

[1]WTTC

The European travel and tourism industry, on the supply side, is very fragmented: 95% of the tourism industry in Europe is composed of very small enterprises. The supply of the tourism product is generated by over a million of these small companies providing a heterogeneous collection of individual travel components (accommodation, attractions, transport, food, events, tours, etc.) that also vary widely in terms of size and scope. The wages and the level of the education of the workers in the tourism supply sector are largely below the average.

The complexity of the tourism value chain and its geographical distribution (with different actors at local, regional, national, global level) has enabled the creation of powerful international and global intermediaries.

Tour operator intermediaries are a driving force of the tourism industry, organising and aggregating the services offered by the supplier and creating all-inclusive standard packages. Small tourism enterprises providing tourism services, notably the ones located in less developed regions, have very little negotiating power faced with these powerful intermediaries on whom they depend for marketing their products. Airlines, hotels, and other travel suppliers are aggressively shifting their focus on-line from providing information to enable direct booking of reservations.

As stated by Pollock and Benjamin[2] "The magnitude of suppliers, intermediaries, distribution channels and consumer segments has bred enormous complexity, disfunctionality and inefficiency into the current system. Pollock separates providers into three categories (as illustrated in Fig. 4.3):

1. Large global and national "bricks and mortar" companies, chains and consortia that have achieved economies of scale and varying degrees of vertical integration in order to be in a position to supply all aspects of a leisure trip or capture the loyalty of a business traveller making frequent and similarly motivated business trips. Corporate examples include BA, Accor, Cendant, Airtours, Hertz etc. Each of these has invested heavily in IT systems to support all aspects of their business;
2. Equally large systems providers and electronic intermediaries that supply a rich blend of content, technical functionality and connectivity — platforms and infrastructures — that are used largely by the first category of provider. Examples include the Global Distribution Systems

[2]Pollock Anna, Benjamin Lean (2002) *Why web services and grid computing will turn the travel and industry on its head — and why that's a good thing!*

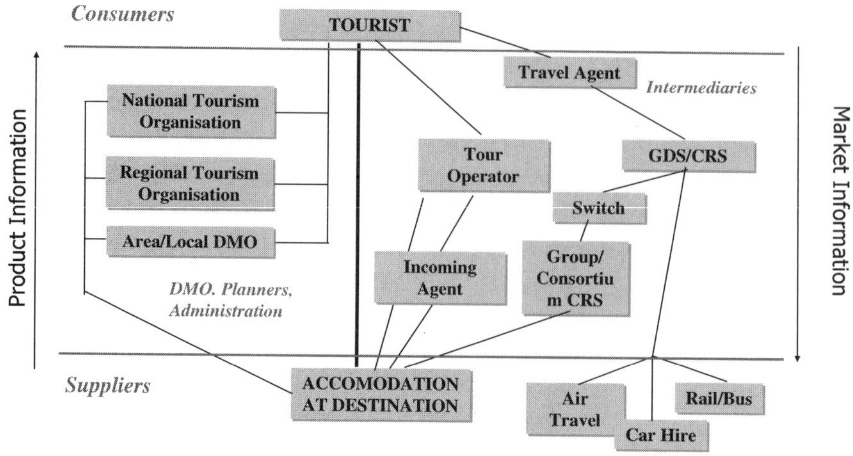

Fig. 4.3. The Tourism value chain.

(GDS) such as Galileo and Sabre, switching companies such as Wizcom and Pegasus and the online agencies of Expedia, Travelocity *et alia*. Corporates in both categories appear to be obsessed with growth, mergers and acquisitions. Cendant has recently purchased Galileo and Cheaptickets and intends to launch its own Internet Travel Portal; Sabre recently announced its intent to acquire 100% ownership of Travelocity; Expedia is championing its Merchant Model — a new name for an old, and supposed beleaguered business concept, the Tour Operator!

3. The rest, hundreds of thousands of SMEs located throughout the globe that offer both direct travel-related services (accommodation, dining, transport, recreation and entertainment) and ancillary services (insurance, software, content, finance, weather, news, maps etc.)".

4.3.2 *The role of information and communication technologies*

As already mentioned, the tourism market is dominated by SMEs that have very limited access to the tourism market because of the lack of distribution channels. ICT, in particular, Internet-based information systems, can help to reduce traditional disadvantages of small tourism suppliers compared to big players. ICT applications within appropriate strategies include:

- Nurturing guest relationships: SMEs have a tradition of building and maintaining relationship to their customers. ICT can support these activities through their provision of computer mediated one-to-one communication allowing an easier maintenance of relationship to a broader range of customers. An established dialogue between service provider and guest helps the provider to find out customers' preferences, in order to further differentiate and tailor offers;
- Building complementary products in one "location" (one-stop-shop);
- Diversifying distribution channels: distribution over the internet provides multiple ways of selling products and services;
- Targeting niche and special interest markets: small providers can react much faster to emerging trends thanks to ICT as their help in this case is providing fast communication with customers.

The travel and tourism industry was one of the forerunners in the implementation of large electronic data exchange networks. Central Reservation Systems (CRS) and Global Distribution Systems (GDS) have been providing on-line distribution, marketing and sale tools to travel professionals for a long time. These systems, initially managing airline reservation, are now addressing more tourism services starting from accommodation. The complexity and the level of the investments created an oligopoly of providers. Sabre, Galileo, Amadeus/START systems serve 60,000 travel agents world world-wide and GDS players are increasingly becoming internet technology providers to the travel and tourism industry, through partnership with business solution developers.

Airlines and travel service providers use GDS to distribute and market their products in real time to travel agencies and airline sales offices connected around the world. These systems are an essential sales tool for travel agencies, which represent the core users.

Travel agencies sell tourism products on behalf of suppliers for a commission and currently rely heavily on Tour Operator packages and airline Global Distribution Systems (GDS). Travel agencies can be seen as "information brokers" but their ability to sell European leisure products is limited by the availability of information about European suppliers ready and able to pay commission — most of which are too small to be included in the databases of GDSs. The merging Internet on-line services, which includes electronic commercial transactions, is revolutionising the traditional distribution marketing channels.

Due to its peculiar structure, the tourism sector is the first sector to be affected by IT and Internet, that will change the way in which tourism

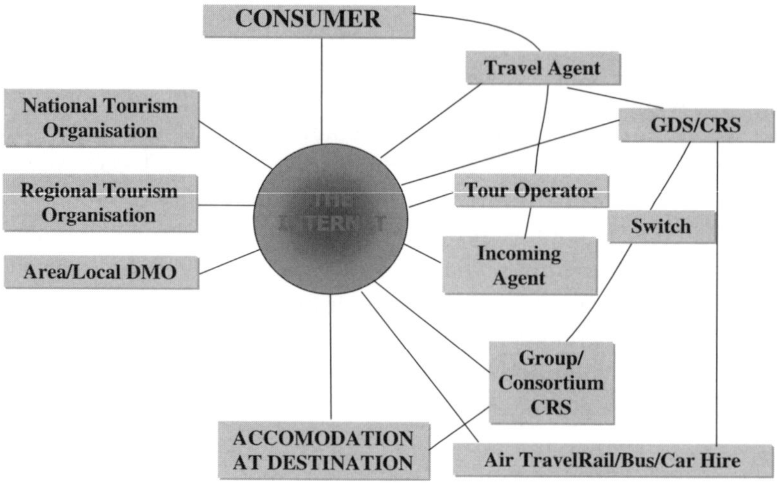

Fig. 4.4. The tourism Internet-based value chain.

will be marketed, booked and sold. Figure 4.4 shows how the Internet can help tourism value chain players to be integrated in a global digital marketplace.

SME operators, however, lack both the capital and the expertise to exploit the global electronic distribution channels and rely on destination marketing organisations (DMOs) or on bigger intermediaries. The investments of DMOs in tourism information and promotion systems are limited and not co-ordinated. Without joining the forces and a policy of cooperation they cannot compete with software and GDS companies.

Within a few years, only a few tourism on-line service providers will continue surviving and only the few entry points that have reached the "critical mass" of information and geographical coverage will remain. This process of concentration, acquisition and creation of alliances will provide to final users, travel agencies and tour operators an acceptable small number of information entry points.

If appropriate measures are not taken early enough, such developments may lead the sector of tourism systems based on the internet to an oligopoly's scenario. Furthermore, non-European companies may well control the European tourism intermediation business. Consequently, the policy, the marketing decisions, the management of tourism flows will very likely not be in the hands of companies which care about the interests of the European regions where the tourism suppliers, enterprises and workers are. The tourism flows will be guided to destinations which can endow higher profits to the intermediaries.

4.3.3 *The next wave of peer-to-peer web services*

As stated by Anna Pollock and Leon Benjamin (2002) "It's the next wave of the Internet — what some have called the Fourth Wave[3] — *peer-to-peer web services* computing that will shake-up this conservative, slow-to-change phenomenon with enormous benefits and implications" specifically for SMEs.

Pollock says that "The Fourth Wave encompasses a number of trends whose interactions are truly synergistic including the move from hard-coded applications to lightweight services delivered over the web; acceptance of fully open standards for data exchange and interoperability; introduction of effective security systems, use of secure micro-payment technologies and peer-to-peer, distributed networks.

Within this wave, it is the widespread use of *"peer-to-peer web services"* (defined as loosely coupled components that communicate via XML interfaces), combined with a *"grid computing model"* (in which every user and his access device is both a client and a server and is both a provider and consumer of information), that will cause a revolutionary upheaval in the way this industry is structured and business conducted".

Looking at the technologies that are enabling the movement towards new models for tourism e-Business we can consider the following as enabling technologies.

1. Pervasive computing using de-personalised devices
   Not only will the Internet be as common as the dial-tone — always on, just there waiting to be used; it will be ubiquitous and accessible from a multiplicity of devices and appliances most of which will not be dependent on wires and landlines for connectivity — Each channel (PC, mobile, interactive digital TV) will play its own unique role. Customers will increasingly expect continuity and consistency of services regardless of the channel used.
2. New tools and standards
   Three major technological advances are enabling this progress towards ubiquity, interoperability and automation: *XML*, *Java*, and *Re-usable* components.

---

[3] *Peer-to-peer web services: Computing's Fourth Wave*, Frank Moss, Bowstreet Software Inc.

- *XML* provides the means whereby the content and structure of the data can be described in a way that machines can interact with it. Think of it this way: while HTML displays, XML does. For example, computer makers might agree on a standard or common way to describe the information about a computer product (processor speed, memory size, and so forth) and then describe the product information format with XML. Such standard way of describing data would enable a user to send an intelligent agent (a program) to each computer makers' web site, gather data, and then make a valid comparison. Any individual or group of individuals or companies that wants to share information in a consistent way can use XML.
- *Java's* Jini network technology is another important innovation that is enabling business to be conducted over the web. It is a programming language that has captured the imagination programmers everywhere because of its ability to "write once, run everywhere". Java's widespread application has done much to speed up the creation and distribution of new and more powerful programmes and applications.
- *Re-usable* components: this important development includes object-oriented programming and use of components. As its simplest, what this means is programmers try wherever possible to re-use software code by identifying the common or generic components that form part of more than one application and applying those components more than once but in different ways to create different applications. What on the surface might appear to be an industry specific application could actually be performed using a generic component: so booking a room of a certain type on a specific date is no different from making an appointment with a certain hairdresser at a salon — both involve a time and space specific event that is sold for a specific price.

3. The arrival of Dynamic Web Services
   Dynamic web services should be dynamic, flexible, interchangeable, network accessible software components that reside on the Internet. They're programming-language agnostic, publish their specific function, can be discovered at design or execution time to participate in a job and interact with other web services. Web services encapsulate business processes (like entering a purchase order or checking a credit card, etc.) and data that can be published in a directory, found by the user (human or machine) and combined with other web services to create more complex applications that can be customised at run time

for each user. At this moment, web services rely on technologies (like UDDI) that do not permit real dynamic connections among them and that is where the new model and technology approach proposed — "OpenMarkets" — comes into its own.

A dynamic web service component is a software "product" that a business makes accessible to customers, suppliers, and partners through the internet. A web service begins as an internal digital asset, such as a software component, an application, a document or a database that provides a set of data or implements a business function.

They will operate between companies, across company firewalls and enable the seamless integration of business between enterprise that, until now, was regarded as impossible. Traditional approaches to B2B integration required months of expensive hard coding. In direct contrast to traditional applications, dynamic peer-to-peer web services provided by OpenMarkets are self contained, self describing and location-independent. This enables them to function autonomously, live everywhere in the network, and be fine-tuned in real time to perform specifically for each individual based on his personal needs, identity and specific characteristics.

4.3.4 The impact of the new business and technological model on the tourism sector

As stated by Anna Pollock (2002) "... looking at the supply-side of travel and tourism, we can see 10 good reasons why Travel and Tourism provides such fertile ground for the application of distributed, peer-to-peer, web services-based solutions:

1. The dominant feature of travel and tourism is supply-side fragmentation. Literally hundreds of thousands of enterprises are involved in providing the various elements associated with overnight travel. Travel/tourism providers are so geographically scattered and numerous that no one technology vendor or distribution channel can dominate.
2. These providers are also characterised by extreme heterogeneity and diversity in terms of focus, size and sophistication.
3. The vast majority are classed as small, medium or micro-sized enterprises. In Europe, some 99% of enterprises employ less than 10 persons. As the vast majority do not have services that can be

purchased online, they are still in a digital sense "invisible" and excluded from e-Commerce.

4. Within each sub-sector and destination, providers actively compete with one another. Cooperation and collaboration only occurs when attempting to reach distant markets and build business volumes to a destination. Autonomy, self-sufficiency and choice are the values and qualities prized by providers.
5. Providers are dependent on a multiplicity of intermediaries, and the distribution chain is long and expensive. They are required to distribute their perishable, time-specific inventory via multiple channels and "brands", managed by intermediaries using a variety of technologies and platforms each requiring separate interfaces.
6. Take-up of sophisticated electronic systems is slow. Most providers will only have dial-up connections. They will not want to join exchanges that rely on private networks or have their data exposed to potential competitors.
7. Providers have neither the financial resources nor the time to manage multiple systems. Fiercely independent, competitive entrepreneurs, they want control over their own inventory management, sales and content.
8. As a result of the volume and fragmentation of both the providers and intermediaries, the industry is awash with thousands of legacy applications that would be costly, if not impossible to dismantle and replace.
9. The sector is particularly vulnerable to the ravages of commoditisation and operates on paper-thin margins. Reducing distribution costs, which can exceed 25% of turnover, provides the best opportunity to improve the bottom line and highlights the importance and advantage of selling directly to the end user.
10. Each provider is embedded in a larger network of suppliers from which it procures the goods and services necessary for operations and a community network of hosts (public and private agencies responsible for helping and benefiting from the economic activity of tourism at the destination)".

On the demand side, there are also a number of powerful factors at work which will encourage the speedy take up of "peer-to-peer web service" technologies over the next decade. Consumer markets are fragmenting into segments of one with each individual expecting higher degrees of personal service tailored to their preferred style and mode of travel.

In addition, the travelling consumer is highly fickle and "event driven". A trip is triggered by a purpose or intent (to visit the sales team; to attend a conference; to go to a wedding; have a weekend break playing golf; or a romantic getaway with partner; or take the annual trip to a favourite cottage by the sea...). Thus the profile of the user changes according to the trip's dimension, focus and rhythm. Each trip is a unique mosaic of supply elements that are experienced by the consumer as a highly subjective, holistic and intangible mental construct called *fantasies* prior to consumption and *memories* afterwards.

Demand factors that will drive providers towards "peer-to-peer web services" may be summarised as:

1. The need to cater to the consumer's *total trip* as a means of demonstrating value, relevance and meaning. In other words, transport carriers and hotels in particular, will need and want to sell a variety of destination services (events, attractions, tours) and ancillary services (books, insurance, weather, finance, news) in order to satisfy their information hungry prospects. Brand loyalty will be built on managing and delivering on the total experience, not just one piece of it.
2. The need to access and offer a broader, richer range of web-based functionality that supports consumers in their trip planning and purchasing tasks; and
3. The need to be visible and accessible in as many places, channels and formats as possible (i.e. by phone, on the company Intranet, corporate web site or third-party agency portal, via wireless phone or PDA, on the back of an airline seat in tantalising digital video, on the TV in the hotel room).

4.3.5 *The new structure of tourism services*

The development of a network of peer-to-peer web services is viewed as a tool for the growth of tourism industries and for the creation of new value-added services (refer to Fig. 4.5). A way to do it is the creation of value added open systems and protocols to allow cooperation among European regions, existing systems and other European actors of the tourism industry. Following this idea, the solution is the creation of a group of federated European information systems, which have to be continuously updated and enriched with new value added services.

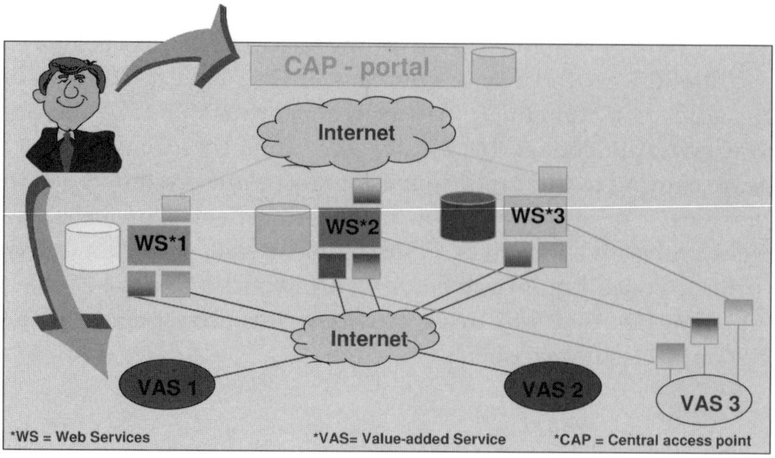

Fig. 4.5. The peer-to-peer web services architecture.

Indeed, the document IST for Tourism[4] highlights the fact that "Europe does not have large software industries, which could afford the investment needed to compete with the American global on-line portal services, and the only model that can cope with this scenario is a federation of open distributed co-operative systems".

In more practical terms, the aim is to establish a federation of information systems and services for tourism (refer to Fig. 4.6). A "federation" is not a new harmonised system, but a set of business agreements, architectural solutions, and interoperability protocols which allow the networking of those information systems that exist already and their interoperability with applications providing value added services.

4.3.6 *The impact of peer-to-peer web services on SME tourism operators*

Peer-to-peer web services will permit the repackaging of existing digital assets (e.g. business processes defined in program code) so they can be wired into other applications. This brings about three profound implications:

[4]IST for Tourism (European Union Report).

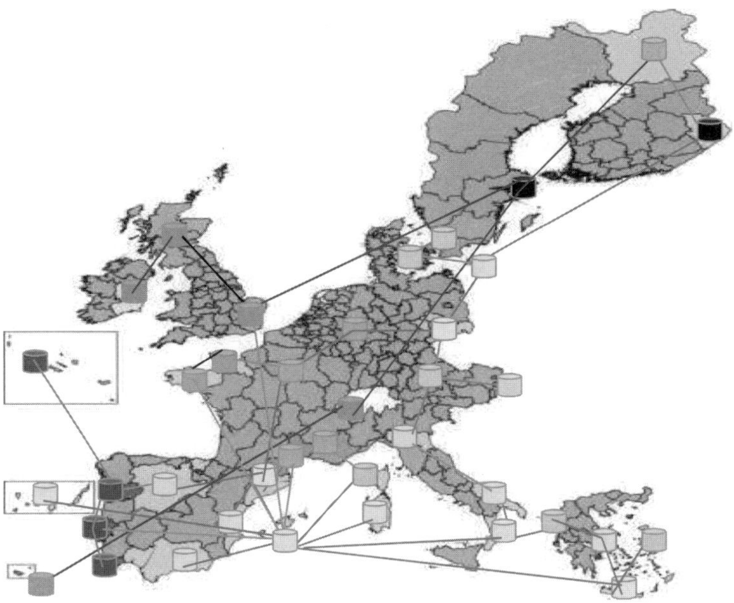

Fig. 4.6. A federated-distributed information system for tourism.

a) Organisations can develop applications by "wiring" peer-to-peer web services together (that are proven, well tested and from trusted sources) thereby decreasing development costs and time to market;
b) Organisations can derive revenue from existing (proven and well tested) code by deploying its business processes to partners in a seamless automated fashion;
c) A technology shift in computing can occur whereby business solutions are constructed from distributed, co-operative application components or services that dramatically reduce and simplify development to enable companies to focus on delivering services relating to their main business, rather than code production.

To understand how peer-to-peer web services and grid computing will change the structure of Travel and Tourism, it is necessary to look at this sector in a different way. The traditional, linear, mechanistic world view that envisions supply chains and the need to own, command and control all the productive units necessary to create and distribute a product is being replaced by more organic, fluid, amorphous and often transitory structures based on alliances, partnerships and collaboration.[5] According to Gartner:[6]

> "The prevailing model of competition in the Internet Economy is more like a web of inter-relationships than the hierarchical, command-and-control model of the Industrial Economy. Unlike the value chain, which rewarded exclusivity, the Internet economy is inclusive and has low barriers to entry. Just like an ecosystem in nature, activity in the Internet Economy is self-organising. The process of natural selection takes place around profit to companies and value to customers."

The end result of widespread deployment of peer-to-peer web services and the service-oriented, web-based architectures on which they are based will be new ways of working and a very different corporate landscape.[7]

4.3.7 *The new tourism ecosystem*

The combination of liberating technologies with use of a new rule book will result in a different business landscape. Instead of power being invested in a relatively small number of very large, vertically integrated organisations, we expect to see a much more dynamic, fluid web of corporate relationships that form, act, dissolve and re-form with enormous speed and flexibility and subject to market demand (refer to Fig. 4.7). Wealth will become more evenly distributed among a much larger number of smaller, more agile enterprises that occupy specific niches in the marketplace and that adapt and evolve quickly and radically according to external conditions. Elsewhere, we have argued that the value chain associated with tourism is better described as a "tourism ecosystem".[8] Following Pollock and Benjamin (Desticorp) full scale deployment of the technologies associated with the Fourth Wave of computing, would force the emergence of five types of corporate "species" in the tourism ecosystem.

---

[5]For a more in-depth exploration of the move to an organic "paradigm", review *Shifting Sands: The Tourism Ecosystem in Transition* at www.desticorp.com/resources.

[6]*Collaborative Commerce*, papers presented by Bruce Bond and Yvonne Genovase, Gartner Symposium, ITXPO 2000.

[7]Anna Pollock.

[8]*Shifting Sands: The Tourism Ecosystem in Transition*, Anna Pollock, www.desticorp.com/ resources.

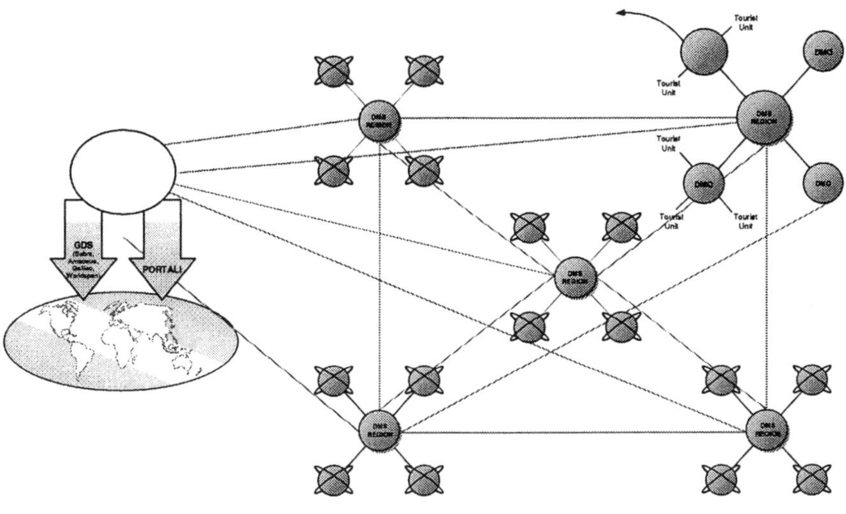

Fig. 4.7. The tourism industry scenario.

A. **The individual peer-to-peer web service provider** — this forms the basic building block of the ecosystem. Equivalent to the atom or the cell, the peer-to-peer web service can comprise either a *unit of information* (description of a provider's facility, a piece of copy describing a place, a video clip) or *unit of functionality* (such as a message to initiate a credit reference check, a currency conversion service, issuance of a booking reference number) or a *unit of content* — combination of content and functionality. Both are wrapped in XML and available over the web. Some companies, such as Sabre or an airline, may offer multiple peer-to-peer web services by disaggregating and repurposing meaningful units of code and functionality. The vast majority will offer one or two (e.g. a self catering holiday cottage, airport taxi service, currency conversion service, a search engine, or credit card checking and processing service).
B. **Brokers** (sometimes known as intermediaries) will aggregate individual services for distribution to customers and their agents. They add value by facilitating easier choice and comparison (helping the customer match their unique needs with the most appropriate supplier) and adding relevant knowledge and expertise. Brokers will tend to be supply-oriented and need to build a critical mass of content in order to offer their customers choice and substance. They will differentiate according to their focus:

- ***Element-focused brokers*** will aggregate specific trip elements.
- ***Destination-focused brokers***: aggregate a variety of content for a specific place (country, region, community or resort). They will also focus on providing services to their members or constituent suppliers in order to access their content and inventory. They will supplement their inventories with rich editorial and interpretative content to aid a consumer's destination choice and appreciation.
- ***Thematic or activity-focused brokers***: aggregate content around a specific topic e.g. walking, cycling, garden viewing, farm holidays, adventure etc. These brokers — many of which are currently no more than directory publishers — will prosper to the extent that they offer relevant knowledge, advice and expertise to add value (e.g. travellers' tips, independent reviews etc.) prior to departure and build and facilitate the exchange of information and reviews among customers.

C. **Channel managers/distributors:** will provide conduits to customers and their agents and will include iTV channels delivered into the home, airline, hotel, office; web portals (wireless and otherwise), Call Centres, print media, kiosks etc.

D. **Affinity agents or life agents:** These are the enterprises that are closest to the consumer and focus all their energy on serving the consumer. It is a symbiotic relationship. They exist to save the customer time and or money and draw down from the web those products and services that the customer needs to solve a problem, complete a task, or meet an objective. In return, the customer entrusts the agent with a substantial amount of personal detail (their personal profile) and encourages the agent to learn more about themselves and their preferences through the act of serving them.

E. **Network Providers** will create and supply the infrastructural glue that binds all participants together and will enable trade to occur in a safe, secure and efficient manner. Over time, all content and functionality (software tools) used by the acts in the tourism ecosystem will be supplied as XML wrapped "peer-to-peer web services" that can be invoked and used over the Internet, assembled to create higher order applications and perform tasks. These services include Security (authentication, authorisation, encryption), Directory, Messaging, Orchestration, Billing & Metering, Quality Assurance).

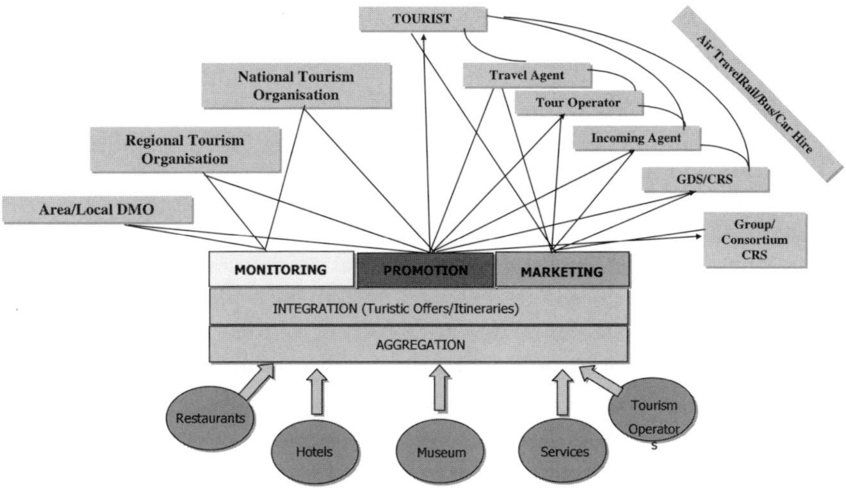

Fig. 4.8. The tourism ecosystem.

4.3.8 *Conclusion*

The tourism industry, no less than many other commercial and financial sectors, needs to develop new ways of working in today's information and knowledge-based Society. This involves developing and validating new types of work-flow and information exchange between suppliers, vendors and customers in the "tourist value chain". An important part of this is the integration and the facilitation of interoperability of existing technologies to promote the use of innovative business methodologies and technologies (refer to Fig. 4.8).

Intermediation plays a key role due to the extreme complexity and fragmentation of the tourism value chain. The deployment of ICTs and the emerging of on-line marketing and information services on the Internet are revolutionising the tourism intermediation affecting the way in which tourism will be marketed, booked, sold. New actors are emerging in the rich market of tourism intermediation.

A strong and effective answer to this process would be the creation of value added open systems, components, system architectures and protocols which allows co-operation among existing systems and the various actors of the tourism industry. In this way a federated peer-to-peer information network could emerge, being continuously updated and enriched with new added value services.

Basically, the on-line services for tourism would be based on a world of wide-distributed network of two basic components:

- Basic (tourism) services: that are information about simple services offered by the suppliers (accommodation, leisure and transport) and available on tourism information systems;
- Value added (tourism) services: which allow access and exploit of the information.

The integration of these services and system in a wide-network should reach a critical mass in order to survive to the global competition. The overall goal would be to turn the local tourism information system and the value added services provided by these systems into a Federated and Distributed System in a way that they would appear to the user as a single system, providing easy access to resources and services.

This new scenario would inevitably contribute to the competitiveness of the European SMEs operating in the tourism and travel sector.

References

Agenda 2010 (1998) *A proposed framework for European action in support of SMEs in tourism. Tourism* **99**(810). The Tourism Society, London.

Archdale G. (1992) A new direction for effective destination marketing and quality, in Yee J. (ed.) Proceedings of the 1992 PATA Destination Database Conference, Singapore, **910**, 13–18.

Arpinar S., Dogac A. and Tatbul N. (2000) An open electronic marketplace through agent-based workflows: MOPPET. *Int. J. on Digital Libraries* **3**(1), 36–59.

Bennett M.M. (1993) IT and travel agency: A customer service perspective. *Tourism Management* **14**(4), 259–266.

Bennett M.M. (1997) Heritage marketing: The role of information technology. *Journal of Vacation Marketing* **3**(3), 272–280.

Binks M. and Gibbs B. (1997) *Beyond Competitiveness: Breaking the Mould: A Case Study of SME Collaboration*. CCPD, University of Nottingham.

Blackburn R., Curran J. and Jarvis R. (1990) Small firms and local networks: Some theoretical and conceptual explorations, National Small Firms Research Conference, Harrogate.

Buhalis D. (1994) ICT as a Strategic tool for small and medium tourism enterprises, in A.V. Seaton *et al. Tourism: The State of the Art*. Wiley, West Sussex, 254–274.

Buhalis D. (1997) Information technology as a strategic tool for economic, social, cultural and environmental benefits enhancement of tourism at destination regions. *Progress in Tourism and Hospitality Research* **3**(1), 71–93.

Buhalis D. (1998) Strategic use of information technologies in the tourism industry. *Tourism Management* **19**(5), 409–421.

Buhalis D. and Cooper C. (1998) Competition or cooperation? Small and medium sized tourism enterprises at the destination, in Laws, E. *et al.* (eds.) *Embracing and Managing Change in Tourism. International Case Studies*. Routledge, London, 324–346.

Camrass, Roger and Farncombe, Martin (2001) *The Atomic Corporation: A Rational Proposal for Uncertain Times*. Capstone.

Chervenak L. (1997) 25 years of hospitality technology. *Bottomline* **12**(4), 13–17.

Clarotti G. (1996) Special measures to allow more SMEs to benefit from European research programmes, globalisation and the SMEs. International Small Business Congress, Athens, November 24–27.

Commission of the European Communities (1996) *Report on the Evaluation of the Community Action Plan to Assist Tourism* 1993–95. Council Decision 92/421 EEC, Brussels, CEE.

Curran J. and Blackburn R. (1994) *Small Firms and Local Economic Networks, The Death of the Local Economy*. Paul Chapman, London.

DATAMONITOR (1997) *New Technologies in Travel and Tourism: Evolving Distribution Channels*. Datamonitor, London.

Destination Management Systems. *Int. Journal of Hospitality Man* **15**(2), 99112.

Ebner A. (1994) TIS tourist information system for the tyrol, in Schertler W. *et al.* (eds.), *ICT in Tourism*: 3542, ENTER '94, Innsbruck, Springer-Verlag, Wein.

Evans G.L. (1997) Networking for growth–SMEs in Tourism and Advanced Producer Services (with Sue Bagwell), SMEs and Networking 4th International Conference, University of Bologna.

Fensel D. (2000) *Ontologies: Silver Bullet for Knowledge Management and Electronic Commerce*. Springer-Verlag, Berlin.

FT (1999) IT Survey, *BusinessZone News*, Financial Times, London, 5th May.

Hagel, John and Singer, Marc (2001) *NetWorth: Shaping Markets When Customers Make the Rules*. Havard Business School Press.

He Q., Sycara K.P. and Finin T.W. (1998) Personal Security Agent: KQML-Based PKK. *Proceedings of the 2nd International Conference on Autonomous Agents* (*Agents-98*), 377–384, New York, ACM Press.

Inkpen G. (1998) *Information Technology for Travel and Tourism*, 2nd edition. Longman, Harlow.

Killdara.com (2001) *Vitiris and the Peer-to-peer web services Ecosystem*, whitepaper.

Laubenheimer M.C. (1999) From intraregional competition towards intraregional cooperation in tourism: The concept of Telecooperation and Virtual Enterprises in the regional tourism business, Paper presented at the Tourist Conference: Tourism Towards the Information Society, Madeira: 19–99.

Laws E. *et al.* (eds.) (1998) *Embracing and Managing Change in Tourism: International Case Studies*. Routledge, London.

Lettmayr C. (1996) *IT as Instrument for Cooperation of SMEs: Globalisation and the SMEs*. International Small Business Congress, Athens, November 24–27.

McAfee A. (2000) The napsterization of B2B. *The Harvard Business Review*, November–December.

McNeill L. (1997) *Travel in the Digital Age*. Bowerdean Publishing Co. Ltd., London.

Mutch A. (1995) IT and small tourism enterprises: A case study of cottage letting agencies. *Tourism Management* **16**(7), 533–539.

Oram A. (ed.) (2001) *Peer-to-Peer: Harnessing the Power of Disruptive Technologies*. O'Reilly.

Passiante G., Elia V. and Massari T. (2001) *Net Economy*. Cacucci Editore.

Pollock, Anna and Benjamin, Leon (2002) *Why Web Services and Grid Computing will Turn the Travel Industry on Its Head — and Why That's a Good Thing!* Desticorp Ltd.

Pollock, Anna (March 2002) *Customers: The G-force that will Pull Web Services into the Frame*. Desticorp Ltd.

Pollock, Anna and Benjamin, Leon (April 2001) *Shifting Sands: The Tourism Ecosystem in Transformation*. Desticorp Ltd.

Porter, Michael E. (2001) Strategy and the Internet. *Harvard Business Review*, 62–78.

Rimmington M. and Kozak M. (1997) Development in information technology: Implications for the tourism industry and tourism marketing. *Anatolia: An International Journal of Tourism and Hospitality Research* **8**, 59–80.

Schertler W., *et al.* (eds.) *ENTER'99 ICT in Tourism*. Springer-Verlag, Wein, Austria.

Sculley A.B. and Woods W.W.A. (1997) *B2B Exchanges: The Killer Application in the Business-to-Business Internet Revolution*. ISI Publications.

Several authors (1998/99) Rural development in the information society. *LEADER II Magazine* **19**, EU, Brussels.

Shackley M. (1997) Saving cultural information: The potential role of digital databases in developing cultural tourism. *Journal of Sustainable Tourism* **5**(3), 244–249.

Sheldon P.J. (1997) *Tourism Information Technology*. CAB International, Wallingford.

Stanworth J., Purdy D and Kirby, D. (1992) *The Management of Success in "Growth Corridors"*. Small Business Research Trust, Open University, Milton Keynes.

Stipanuk D.M. (1993) Tourism and technology: Interactions and implications. *Tourism Management* **14**(4), 267–278.

Strategic Advisory Group (1997) Information Society Technologies (IST) for Tourism, Report on the 5th Framework program on Information Society applications for transport and associated services "Think Tank for Tourism", 8th December.

Sussmann S. and Baker M. (1996) Responding to the electronic marketplace: Lessons from destination management systems. *Int. J. Hospitality Management* **15**(2), 99–112.

TIA (1997) *Travel and Interactive Technology: A Five Year Outlook*. Travel Industry Association of America, Washington, D.C.

Tsalgatidou A. and Veijalainen J. (2000) Challenges in Mobile Electronic Commerce. *Proceedings of IeC 2000, 3rd International Conference on Innovation through E-Commerce*. Manchester, UK, November 14th–16th.

Van der Borg J. *et al.* (1998) *La diffusione delle tecnologie informatiche e telematiche nell' industria turistica italiana*. CISET, Venezia.

Vickerstaff S. and Parker K.T. (1995) Helping small firms: The contribution of TECs and LEC's. *International Small Business Journal* **13**(4), (1995).

Werthner H. and Klein S. (1999) *Information Technology and Tourism — A Challenging Relationship*. Institute for Information Systems, University of Muenster, Germany.

Wise R. and Morrison D. (2000) Beyond the exchange: The future of B2B. *Harvard Business Review*, Nov–Dec 86–96.

WTO (1998) Tourism and Technology. *Bulletin*, **3**, WTO, BC, Madrid.

5

Knowledge Leadership to Drive Digital Innovation

Giuseppina Passiante and Valerio Elia

5.1 Introduction

The phenomena presented in the previous chapters suggest that a new model of leadership needs to be identified, which would guide innovation in any organisation competing in the Digital Economy, including both firms and regions. Indeed, our vision is based on the metaphor of "a region as an enterprise".

Recent studies have identified a new "leadership paradigm", as the capability to enhance the Intellectual Capital of an organisation (Senge, 2000; Sveiby, 2000). This aim is achieved through a continuous and dynamic development, enhancement, and use of the three components of Intellectual Capital: Human, Structural, and Social Capital. But this aim is also achieved through an organisation-wide integration of the three components. This dynamic framework, in which Intellectual Capital plays a fundamental role, suggests a rethink of the knowledge management concept in terms of "Knowledge Leadership". Indeed "leadership" is about constant development and innovation, so it is a better term than "management".

In this chapter, the following points are discussed:

a. The emerging competitive environment of the Digital Economy, which gives an overview of its characteristics;

b. A re-thinking of conventional theories, which have been unable to provide a holistic framework for value creation in the Digital Economy;
c. A new concept of innovation, in the new competitive landscape. This Digital Innovation highlights the need for a new "Leadership paradigm";
d. "Digital Innovation Leadership" as "Knowledge Leadership";
e. The "organisational knowledge creation" process as a fundamental process driven by Knowledge Leadership;
f. A case study concerning the creation of Knowledge Leadership for digital innovation.

5.2 The Emerging Competitive Environment of the Digital Economy

A wide body of literature has analysed the Internet-driven competitive environment in the Digital Economy (Hagel and Rayport, 1997; Upton and McAfee, 1996; Armstrong and Hagel, 1996; McKenna, 1995; Handy, 1995; Davis and Botkin, 1994; Moore, 1993). These studies have highlighted some stylised facts that emerge from the analysis of this new competitive environment (Elia and Passiante, 2001):

- The flourishing of "sense and respond organisations";
- The development of value networks, focused on customer's needs, where competition and coevolution processes coexist.

5.2.1 *The flourishing of "sense and respond organisations"*

Unpredictable, discontinuous change is an unavoidable consequence of the Digital Economy. This environment demands fast and sometimes instantaneous response, and as a result many organisations are fragmenting themselves into smaller units in order to respond in real time (Evans and Wurster, 1997).

In this ever-evolving environment, organisation leaders need to be able to change and adopt new organisation models instantaneously. Organisations must embrace not only *adaptive mindsets but also adaptive organisation platforms* (Haeckel and Nelson, 1993; Haeckel, 1995). These platforms are tied together by humans collaborating to deliver a "sense and respond organisation", also known as an "event driven organisation".

This new view of the organisation is forcing a re-examination of the assumptions and techniques used to support traditional organisation functions such as communication, evaluation, negotiation, coordination and resolution making. These basic components must be present for a complete architecture, but agility and adaptation, simplicity and focus are now vital attributes for people and systems in the new ecology of organisation. How quickly and efficiently an organisation can respond to change depends upon what is known about the environment and what is done about it.

The responsiveness of an evolving organisation is a reflection of its collaborative environment (Bessant and Tsekouras, 2001). With a simple, adaptive, holistic and contextual solution, organisations have the ability to offer value and service to all stakeholders.

Collaboration is typically defined as "working jointly with others, especially in an intellectual endeavour". In order to truly collaborate, shared meaning, processes and commitment must exist to create an "intellectual endeavour" (Nishiguchi, 2001). This concept of sharing leads to a new wisdom based on the collective capabilities of an organisation.

Today's market leaders demand tightly integrated relationships with their suppliers, customers and partners in order to speed innovation, build lasting relationships, and increase productivity and speed profitability. The primary driver of growth and value of these organisations is their ability to collaborate across these relationships.

Sense-and-respond organisations require a strategic, collaborative and "all stakeholder inclusive" decision-making process in order to be successful in today's turbulent organisational arena.

A key role is played by customers. Indeed, in the Digital Economy, the gap between producers and consumers is blurring. As highly customised products and services replace mass production, producers must create specific products that are imbued with the knowledge requirements, and tastes of individual customers. In the Digital Economy, consumers become involved in the actual design process (Reicheld and Schefter, 2000; Iansiti and MacCormack, 1997). This perspective of blending production and consumption is reminiscent of the notion of "prosumption" suggested by Alvin Toffler.

Customers, not just suppliers, increasingly create value by drenching products with their knowledge; customers can disaggregate value and re-aggregate it from multiple suppliers enhancing their power. Therefore the emphasis is on customer relationships, as opposed to brand images.

5.2.2 *The development of value networks*

In the Digital Economy, instead of value chains, value networks are emerging (Amitt and Zott, 2001). These value networks are multi-organisation networks of relationships focusing on the integration of information flow (Tapscott, Ticoll, and Lowy, 2000). The purpose of the value chain is to exploit information and knowledge within the organisation. However, the fundamental unit of the Digital Economy is not the organisation but the individual. Tasks are not assigned and controlled through a stable chain of management but rather are carried out autonomously by independent contractors. These electronically connected freelancers, e-lancers, join together into fluid and temporary networks to produce and sell goods and services. When the job is done, after a day, a month, a year, the network dissolves, and its members become independent agents again, circulating in the economy, seeking the next assignment. All these trends point to the devolution of large, permanent corporations into flexible, temporary networks of individuals.

The successful organisation of today has transformed from classic, closely held bureaucracy to a centre-less entity that can spread across the globe. The centre-less organisation accomplishes its mission using a rich set of evolving relationships with its stakeholders, both internally (employees, managers, board of directors) and externally (partners, suppliers, customers) under one common identity (Malone and Laubacher, 1998).

At a time when the value chain concept is giving way to value networks, the emphasis needs to shift from minimising transaction costs incurred by single organisations to maximising transactional value created by networks of organisations (Amit and Zott, 2001).

In this context, an organisation is viewed as a part of an ecosystem that crosses a variety of industries. In an ecosystem, organisations "co-evolve" capabilities around a new innovation, working cooperatively and competitively to support new products and satisfy customer needs. An organisation ecosystem, like its biological counterpart, gradually moves from a random collection of elements to a more structured community.

In this sense, the world of nature, in all its complexity, offers rich potential for gaining a deeper understanding of the fundamental dynamics of the equally complex world of organisation. They are organic systems open to their environment, creatively evolving and self-organising in order to renew themselves while maintaining internal coherence.

Such organisations focus not only on "how to be a good competitor" but also on "how to be a good evolver". Long-term superior performance is achieved through continuously developing and adapting new sources of temporary advantages.

5.3 Re-thinking the Conventional Theories

The phenomena emerging from the observation of the new competitive environment, represent a profound change in the ways organisations operate and in how economic exchanges are structured. They also open up new opportunities for wealth creation, and represent a challenge to the conventional theories of how value is created in a single firm or in a region. In the following and non-exhaustive review of the literature, some of the main contributions focused on single firms and regions are presented.

5.3.1 *Conventional theories about business strategic management*

Several phenomena emerging from the observation of the new competitive environment represent a profound change in the ways organisations operate and in how economic exchanges are structured. They also open up new opportunities for wealth creation, and represent a challenge to the conventional theories of how value is created (Amit and Zott, 2001).

According to Amit and Zott (2001) there are several conventional theoretical frameworks that make valuable suggestions about possible sources of value creation in a firm. In the following paragraphs, we will give a brief summary of the main theories.

Value chain analysis

Porter's value chain framework (1985) analyses value creation at firm level. According to Porter, value creation cannot be understood by looking at a firm as a whole. Value chain analysis identifies the many discrete activities a firm performs in designing, producing, marketing, delivering, and supporting its product. Value can be created by differentiation along every step of the value chain, through activities resulting in products and services that lower buyer's costs or raise buyers' performance.

In the Digital Economy the value creation opportunities may result from new combinations of information, physical products and services, innovative configurations of transactions, and the reconfiguration and integration of resources, capabilities, roles and relationships among suppliers, partners and customers, that the value chain analysis fails to capture (Amit and Zott, 2001).

Schumpeterian innovation

Schumpeter (1934) considers innovation as the source of value creation. According to Schumpeter, technology and novel combinations of resources are the foundations of new products and production methods. These, in turn, lead to the transformation of markets and industries, and hence to economic development. The Digital Economy has amplified the notion of Schumpeterian innovation, since innovation spans firm and industry boundaries, involves new exchange mechanisms and unique transaction methods (rather than merely new products, or productions processes) and fosters new forms of collaborations among firms.

Resource-based view of the firm

The resource-based approach considers the firm as a bundle of resources and capabilities that may lead to value creation (Penrose, 1959; Wernerfelt, 1984; Prahalad and Hamel, 1990; Barney, 1991; Peteraf, 1993; Amit and Schoemaker, 1993). A firm's resources and capabilities are valuable if, and only if, they reduce a firm's costs or increase its revenues compared to what would have been the case if the firm did not possess those resources. A more recent approach, the dynamic capabilities approach (Teece, Pisano and Shuen, 1997), explores how valuable resource positions are built and acquired over time.

The Digital Economy presents a challenge for the resource-based approach because as the information-based resources and capabilities, which have a higher degree of mobility than other types of resources and capabilities, increase in their importance within firms, value migration is likely to increase and the sustainability of newly created value may be reduced. The prospect of value preservation or sustainability is an important incentive for value creation (Amit and Zott, 2001).

Strategic networks

The strategic networks approach focuses on the implications of strategic networks (stable inter-organisational ties which are strategically important to participating firms) for value creation. Strategic networks may take the form of strategic alliances, joint ventures, long-term buyer-supplier partnerships, and other ties (Gulati, Nohria and Zaheer, 2000).

There are several sources of value creation in strategic networks: trust, resources and capabilities (especially those of suppliers and customers), economies of scale and scope, knowledge sharing, shortened time-to-market, enhanced transactions efficiency, reduced asymmetries of information, and improved coordination between the firms involved in an alliance.

However, strategic networks theory only partially explains the value creation potential of firms based on new business models. Indeed, the Digital Economy, with its unprecedented reach, connectivity, and low-cost information processing power, opens up entirely new possibilities for value creation through the structuring of transactions in novel ways. These new transaction structures are not fully captured by network theory (Amit and Zott, 2001).

Transaction cost economics

Transaction cost economics identifies transaction efficiency as a major source of value, as enhanced efficiency reduces costs. It suggests that value creation can derive from the attenuation of uncertainty, complexity, and information asymmetry (Williamson, 1975). The Digital Economy impact on firms and organisations is the reduction of coordination costs and transaction risks (Amit and Zott, 2001). In addition to decreasing these direct costs of economic transactions, Information and Communication Technologies also reduce indirect costs, such as the costs of adverse selection, moral hazard, and hold-up. Nonetheless, the emphasis of transaction cost economics on efficiency may divert attention from other fundamental sources of value such as innovation and the reconfiguration of resources. The theory also focuses on cost minimisation by single parties and neglects the interdependence between exchange parties and the opportunities for joint value maximisation that this presents (Amit and Zott, 2001).

The limits of each one of the previous approaches highlight the necessity to rethink the conceptual framework of business strategic management. Indeed, the Digital Economy is generating a continuous change, altering the fundamental ways organisations conduct business to create wealth. Change is constant, and is creating a competitive landscape with substantial uncertainty. Because of this uncertainty, scholars are trying to define new theoretical frameworks, where strategic management research integrates an entrepreneurial perspective, where entrepreneurship is conceived as the identification end exploitation of previous unexplored opportunities (Amit and Zott, 2001). Entrepreneurial actions entail creating new resources and combining existing resources in new ways to develop and commercialise new products, move into new markets and/or serve new customers.

On the other hand, strategic management entails a set of commitments, decisions and actions designed and executed to produce a competitive advantage and above-average returns. Strategic management calls for choices to be made between competing alternatives.

"On the edge" research aims at integrating entrepreneurial and strategic perspectives, to examine the entrepreneurial strategies that create wealth. These new approaches, called "strategic entrepreneurship" approaches, are a way to face the challenges of the Digital Economy.

5.3.2 *Conventional theories about competition and location*

In Chapter 1, we described some seminal approaches that are mainly focused on the knowledge exchanges among the partners of the regional clusters, in order to gain some insights to help compare the traditional role of geographic proximity in generating value for a geographic cluster with the analogous role of the virtual proximity in the Digital Economy. The list included:

- Marshall-based approaches, that highlight the role of local social capital, as mainly pointed out by the "industrial districts" approach;
- Porter's approach, centred on the dynamics of geographic clusters in supporting innovation processes and competitiveness;
- The evolutionary approach, that emphasises the role played by local knowledge spillovers and collective learning processes, mainly developed in the "*milieu innovateur*" approach.

In all these approaches, the nature of clusters and their value are related to the issues that historically characterise the interrelationships existing between the economic organisation and the spatial change of an economy.

The emerging Digital Economy is creating new interrelationships between economic organisation and spatial change, related to the digital innovation paradigm. In this economic landscape, a shift is emerging from geographical industrial clusters to virtual clusters, driven by digital innovation. The industrial clusters that are emerging in the Web-based world of business highlight a new interrelationship between economic and spatial organisation, a new competitive space where "how you do business" is more relevant than "where you do business".

Internet-based digital connectivity, is creating new interrelationships between economic organisation and spatial change: in this way, new patterns of economic activities are emerging. The new Internet-based "economic organisation" is determining different forms of spatial organisation of the industrial clusters. The positive interference between organisational proximity and geographical proximity is generating new link-spaces, that are characterised by ubiquitous provision and distribution of information in networks of agents. "Virtualness" becomes a relevant source of value, because it enhances the advantages coming from the multidimensional character of proximity, and introduces new ways and mechanisms of specialisation and integration and exchange of knowledge.

The basis of competitiveness comes from network organisations, mutually dependent relationships, globally oriented physical and communication infrastructure, knowledge creation and continuous improvement processes. These stylised facts highlight the necessity to rethink the role of local/regional/national governments in facilitating the increasingly important role of "created assets", which are more suitable for the new clusterisation forms, and the characteristics of the leaders who have to run these institutions.

5.4 Digital Innovation Calls for Knowledge Leadership

In the Digital Economy some specific characteristics of the innovation process are emerging, both in enterprises and in regional innovation processes. These new characteristics may be identified mainly in:

- Very high-velocity interactions;
- Very rich potential connectivity, involving many different players;
- Global orientation, where distance has become less relevant;
- Pervasiveness of innovation. Innovation is ubiquitous in every industry, in every place, in every firm. For Janszen (Janszen, 2000), "after efficiency, quality and flexibility, we now live in the age of innovation".

Given these characteristics of innovation, the key strategic challenge is managing that innovation. The dilemma is how to do it.

At one level, managing innovation means reacting to change, coming up with better products, creating services that exploit innovation, repackaging existing products. But managing innovation also means anticipating change, lining up resources (e.g. venture partners, cross cultural employees, currency trading skills) early, developing corresponding marketing channels, creating technical options.

Finally, at the highest level, managing innovation is about leading innovation, creating new technologies and products, launching new markets, raising industry standards, redefining customer expectations, increasing the pace of industry product cycles. The strategic challenge is therefore to manage innovation by reacting when necessary, anticipating wherever possible and leading innovation when the circumstances are right (Elia and Passiante, 2001).

The main challenge for innovation management is then related to the organisation's capacity to sense opportunities and to reconfigure itself accordingly. It is what Teece, Pisano, and Shuen (1997), refer to as "dynamic capabilities". *In this sense the capacity of digital innovation is a dynamic capability* (Elia and Passiante, 2001).

In the digital innovation environment, characterised as a high-velocity market, a dynamic capability is a simple, experiential, unstable process, that relies on quickly created new knowledge (Eisenhardt and Martin, 2000). In a high-velocity market, according to Eisenhardt and Martin, uncertainty cannot be modelled as a probability, because it is not possible to identify possible future states, so a dynamic capability necessarily has to rely much less on existing knowledge and much more on rapidly creating situation-specific new knowledge. To allow new knowledge to be rapidly acquired in each new situation, it is necessary to carry on experiential activities, such as prototyping real time information, multiple options, and experimentation, generating new knowledge that quickly replaces existing knowledge (Eisenhardt and Martin, 2000).

Skyrme has identified seven strategic types of knowledge:

1. Customer knowledge. This is the organisation's most vital knowledge;
2. Knowledge in people. Knowledge in people's heads;
3. Knowledge in products and services. Knowledge embedded in products and enhanced knowledge-intensive services;
4. Knowledge in processes. Knowledge embedded in organisation processes and in management decisions;
5. Organisational memory. It helps to identify "knowing what we know";

6. Knowledge in relationship. Improving knowledge flows between suppliers, employees, shareholders, community, etc.;
7. Knowledge assets. Identifying intellectual and knowledge assets

These different types of knowledge are generally grouped into what is called the "Intellectual Capital" of an organisation, classified according to three dimensions (Seemann *et al.*, 2000):

1. *Human Capital* is the knowledge, skills, and experiences possessed by individuals inside an organisation. It comprises both tacit and explicit conceptual knowledge;
2. *Structural Capital* includes the explicit knowledge embedded in the organisation's work processes and systems, or encoded in written policies, training documentation, or shared databases of "best practices". It also includes intellectual property, i.e. patents and copyrights;
3. *Social Capital* is reflected in the ability of groups to collaborate and to work together. Social Capital is the overall set of relationships that make organisations work effectively (Prusak and Cohen, 2001). Often, Intellectual Capital approaches fail because of this critical component. Indeed Social Capital is a function of trust and cooperation. The management's efforts should be to enhance trust and to foster cooperation among the members of the organisation.

In the digital innovation context, leaders thus need to regard Intellectual Capital as a strategic asset. To enhance Intellectual Capital and to ensure that it generates superior organisation results, leaders need to start and carry on a continuous and dynamic development, enhancement, and use of the single elements of Intellectual Capital, but also through an organisation-wide integration of Human, Structural and Social Capital (Romano, 2001).

In this view, Intellectual Capital becomes "a capacity for effective action" (Senge, 2000), not a thing. However, the word "management" suggests custodianship, even control, and a concentration on managing resources that already exist. Remembering Skyrme's concept of Leadership in terms of constant development and innovation, a better definition is "Intellectual Capital Leadership" or more directly "Knowledge Leadership". Knowledge Leadership embraces both the sharing of what is known, and innovation. According to Skyrme, Knowledge Leadership provides a metaphor for the future that goes well beyond the immediate fad of knowledge management.

5.5 Digital Innovation Processes Driven by Knowledge Leadership

Having identified Digital Innovation Leadership with Knowledge Leadership, the question is: what process for creating and sustaining Intellectual Capital is most suitable for digital innovation?

The answer to this question involves a preliminary comprehension of the basic processes for knowledge creation and knowledge sharing.

In this direction a useful operational framework may be "Organisational Knowledge Creation", as proposed by Nonaka and Takeuchi (1995), and developed by Nonaka, Toyama and Konno (2001). Nonaka, Toyama and Konno propose a model of Organisational Knowledge Creation process that consists of three elements:

- The process of knowledge creation via conversion from tacit to explicit knowledge (SECI process). The SECI process is characterised by two dimensions:
  - An epistemological dimension, based on Michael Polanyi's (1966) distinction between tacit knowledge and explicit knowledge. Tacit knowledge is personal, context-specific, and therefore hard to formalise and communicate. Explicit or "codified" knowledge refers to knowledge that is transmittable in formal, systematic language;
  - An ontological dimension that describes the "organisational" amplification of the knowledge created by individuals. This process takes place within an expanding "community of interaction" which crosses intra- and inter-organisational levels and boundaries.

  In this two-dimensional space, the tacit-explicit interaction processes generate a "spiral" of organisational knowledge creation, in which the mobilised tacit knowledge created and accumulated at individual level is "organisationally" amplified at higher ontological levels.
- The shared context for knowledge creation (Ba). As highlighted in Chapter 1, Ba is a "shared context", where knowledge is created, shared and diffused. The key concept that underlines "ba" is "interaction". Knowledge generation is considered a dynamic process that develops through interactions among individuals or between individuals and their learning environment. Ba is then considered as the context shared by those who interact with each other and, via such interactions, create and share knowledge. "Ba has a complicated and ever changing nature. It is a boundary for interactions among individuals, yet its boundary is

open.... Ba is an open place where participants with their own contexts can come and go and the shared context (that is Ba) can continuously evolve. By providing a shared context in motion, Ba sets binding conditions for the participants by limiting the way in which the participants view the world. Yet, it also provides participants with higher viewpoints that their own" (Nonaka, Toyama and Konno, 2001). "Ba" is thus framed as a platform of knowledge that provides the energy, quality and place to perform the individual's/group's/organisation's learning processes.

- The inputs, outputs and moderators of the knowledge creation process (called Knowledge Assets). Knowledge Assets have been grouped into four classes:

  - Experiential knowledge assets, that consist of tacit knowledge shared through common experiences, such as the skills and know-how of individuals, and emotional knowledge (care, trust, security, passion, tension). Since these assets are mainly tacit, they are difficult to grasp, evaluate or trade;
  - Conceptual knowledge assets, that consist of explicit knowledge articulated via images, symbols and language, based on concepts held by members of the organisation, such as products/services, concepts;
  - Systemic knowledge assets, i.e. systematised and packaged explicit knowledge, such as explicitly stated technologies, product specifications, manuals and documented and packaged information (documents, manuals, database) ;
  - Routine knowledge assets, consisting of knowledge that is routinised and embedded in the actions and practices of the organisation (organisational routines and culture).

  The assets become the basis of the knowledge creating process: they are dynamic, and new knowledge assets are created continuously from existing ones.

The three elements of the knowledge creation process interact with each other (Fig. 5.1): using existing knowledge assets, an organisation creates new knowledge through the SECI process, enabled by the organisation's Ba. The knowledge created then increases the knowledge assets of the organisation, that again is the basis for a new spiral of knowledge creation.

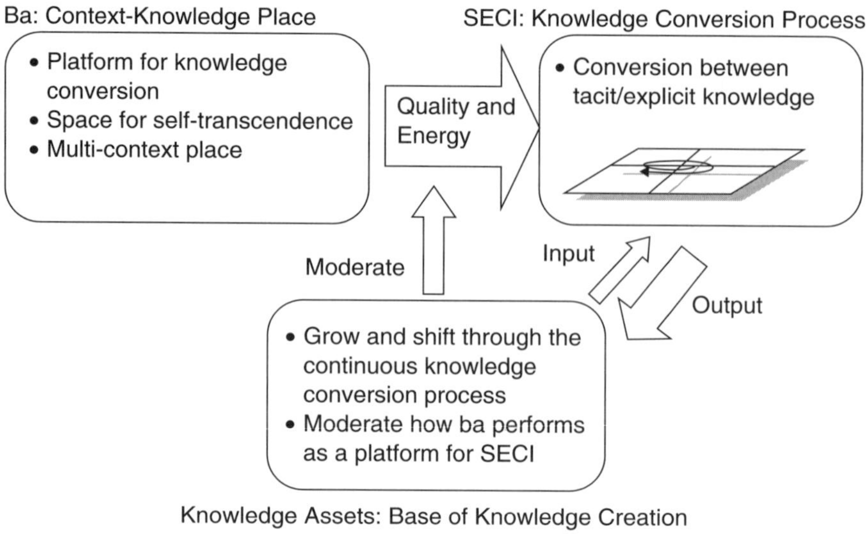

Fig. 5.1. Three elements of knowledge creation (Nonaka, Toyama and Konno, 2001).

In this view, the organisational knowledge creation process needs an enabling context to take place. In order to foster the organisational knowledge creation process of an organisation, Nonaka, Toyama and Konno propose the following actions (Fig. 5.2):

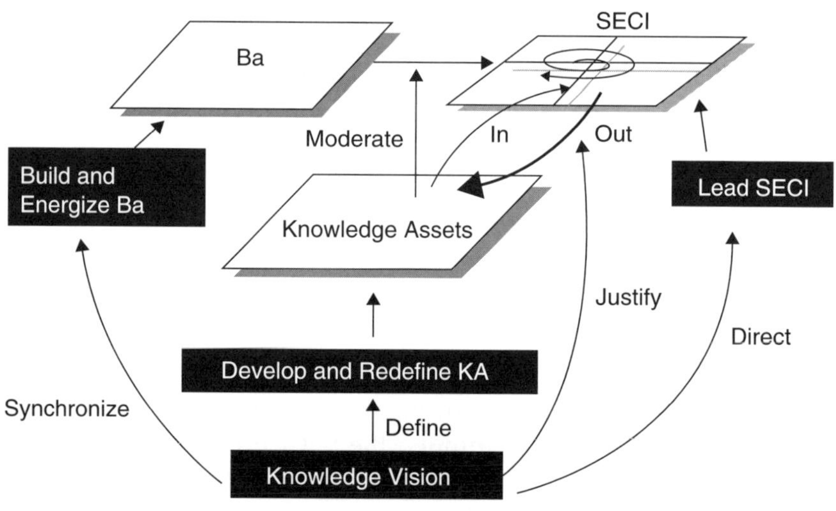

Fig. 5.2. Leading knowledge creating process (Nonaka, Toyama and Konno, 2001).

- Providing the knowledge vision;
- Promoting the SECI process;
- Developing and promoting the sharing of knowledge assets;
- Building, connecting and emerging Ba.

As concerns knowledge vision, Nonaka, Toyama and Konno highlight the need for a vision to synchronise the entire organisation. Leaders have to articulate the knowledge vision and communicate it inside and outside the organisation. The knowledge vision also include the value system of the organisation that, together with the organisational norms and routines, determines the kind of knowledge that has to be created, integrated and exploited.

Following the directions provided by knowledge vision, leaders have to foster the SECI process by facilitating all the four modes of knowledge conversion. Nonaka and Takeuchi's model postulates four different modes of knowledge conversion (Fig. 5.3):

- The socialisation process, that allows the individual's tacit knowledge to be transferred to others, through observation, imitation and practice;
- The externalisation process, where individuals try to articulate knowledge into explicit knowledge, taking the shape of metaphors, concepts, hypotheses or models;

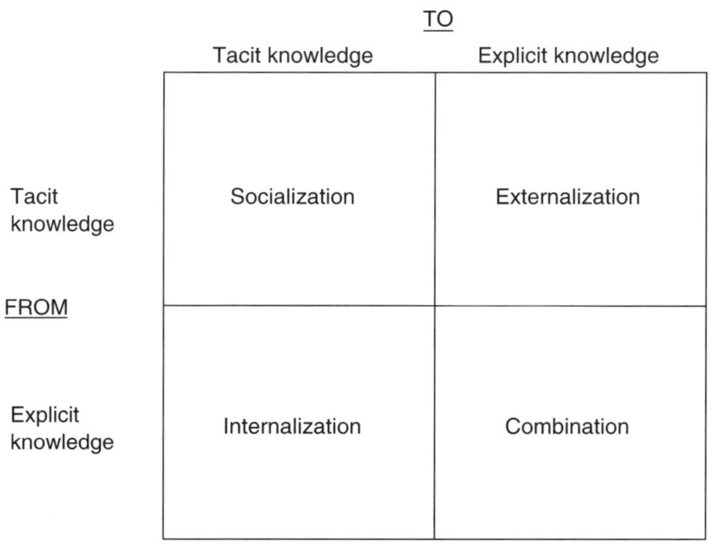

Fig. 5.3. Four modes of knowledge conversion (Nonaka and Takeuchi, 1995).

- The combination process, characterised by individuals exchanging and combining explicit knowledge through such media as documents, meetings, telephone conversations, computerised communication networks, so as to create new explicit knowledge;
- The internalisation process, through which experiences related to socialisation, externalisation and combination processes are internalised into an individual's tacit knowledge base, in the form of shared mental models, or technical skills.

These processes are facilitated by:

- Comprehending the current situation, in terms of where the spiral is heading and what kind of knowledge may be converted;
- Realising the necessary changes in the knowledge creation process;
- Facilitating the use of a common language.

As regards the development and sharing of knowledge assets, Nonaka, Toyama and Konno suggest leaders to determine what kinds of knowledge assets are available to them, and, on this basis, to define a strategy to build, maintain and exploit these assets.

As for building, connecting and emerging Ba, Nonaka, Toyama and Konno stress that Ba can be built intentionally or created spontaneously. However, knowledge participants need some necessary conditions, such as:

- Autonomy, that increases the commitment of individuals to create new knowledge, and may be the source of unexpected knowledge;
- Creative chaos, that encourages people to transcend existing boundaries to define and solve problems;
- Intentional overlapping of information about business activities, to support a sharing of tacit knowledge and the definition of single members' roles;
- Diversity, in order to deal with the challenges of the complexity and the variety of the environment;
- Trust and commitment, since interactions among individuals participating in a Ba or between different Ba has to be supported by trustful sharing of knowledge and continuous exchanges between all units of the Ba.

The creation of a suitable context, enabling organisational knowledge creation, may concern not only a single firm, but also a region: indeed Nonaka, Toyama and Konno conclude their chapter by pointing out that "if we further raise the level of analysis, we arrive at a discussion of how so called national systems of innovation can be built. For the immediate future, it will be important to examine how organisations, governments and universities can work together to make knowledge creation possible" (Nonaka, Toyama and Konno, 2001).

A crucial dilemma then becomes: how is it possible to create leaders who are able to manage this process, and then to foster digital innovation processes in a firm or in a region? As emerge from the previous paragraphs, Knowledge Leaders need to have some specific capacities:

- A shift of competencies from a focus on economic performance indicators (market share, sales, profit, Return On Investments, Cash flow) towards measures, related to knowledge management (people competencies, IT systems and databases, learning capabilities, management capabilities, intellectual property, management philosophy);
- A capacity to manage Human Capital, which is becoming a critical component of Intellectual Capital, because the creation of Human Capital is a complex process, generating a wide variety of results: experiences, know-how, skills, creativity;
- A capacity to scan across different disciplines and industries, that is the key to see emerging conditions, paradigm shifts and opportunities from the competitive landscape of the Net Economy. This new openness creates opportunities for synergistic thinking and cooperative research, focused on the impact of ICTs, on the rise of knowledge as a key economic resource, on the changing face of competition, on global market dynamics, on government-driven changes. These issues provide an opportunity to rethink, redirect, re-organise, re-position and re-build the theoretical and practical business framework;
- A capacity to look at the whole system, not just at the parts; the whole is a constantly changing field of connections, relationships and patterns of interactions. Innovation occurs as the result of new influences that disrupt existing patterns; out of these disruptions, new patterns emerge, seen as issues, trends and technological developments. Non-linear thinking is critical to recognise these patterns, because they are the result of relationships between order and disorder states;
- A capacity to create and use a strategic intelligence network for both qualitative and quantitative information. Information and Communication Technologies (ICTs) have been dramatically changing the way organisations work. Non-linear innovation will need highly efficient electronic markets for ideas, capital and talents.

These capacities may be considered basic concepts and principles to experiment an incubator of Knowledge Leadership, able to create organisations skilled in creating, acquiring and transferring new knowledge and insights.

5.6 Creating Knowledge Leadership for Digital Innovation: The eBMS Experiment

Based on the above guidelines, we have defined a model of "Knowledge Leadership Incubator" that characterises the "e-Business Management School" (eBMS), an experiment that we are currently undertaking at the ISUFI-University of Lecce (Italy).

In a previous book, "Creating Business Innovation Leadership", the eBMS model is described in detail (Romano, Passiante and Elia, 2001). Here, some main aspects of the eBMS experiment are recalled, concerning:

- The eBMS strategy, that is its major goals and actions in time and space;
- The eBMS organisation, including:
  - The eBMS organisational structure and processes;
  - The eBMS resources, in terms of distinctive competencies, and physical/informational resources;
- The eBMS environment, that is the network of relationships that it has developed;
- The eBMS performance, in terms of the value generated by the experiment.

5.6.1 *The eBMS strategy*

eBMS aims at creating a "microcosm of real settings", conceived as a Knowledge Leadership Incubator, where talented people can learn how to learn together, enhancing their collective intelligence, competence and capacity to imagine research questions and pilot projects oriented to generate innovation in the Digital Economy environment.

The value proposition and strategic intent are the creation of Human Capital, conceived as a relevant driver for generating Structural Capital and Social Capital in the target organisations.

eBMS targets are talented people and executive managers, coming from Italy and from Euro-Mediterranean Countries. The target employers of this Human Capital are mainly organisations involved in cultural change programs, in knowledge management and e-Business projects, but also public agencies, research centres/universities and private venture capitalists engaged in creating/generating new business.

The basis for differentiation is grounded on the cross-cultural learning strategies and learning processes, which integrate experiential knowledge, emerging knowledge and formalised knowledge, so as to create the capacity to boost Old Economy — Digital Economy fertilisation. This fertilisation is a complex process, involving many management issues that need cultural change and capacity to scan across different disciplines and industries. In order to face this complexity, the experiment is boosted by strong partnerships and strategic alliances with academic and business actors, at the national and international level.

This strong partnership enables a continual movement between practice and theory, and nurtures the combination of meaningful organisation issues with meaningful inter-organisational dynamics.

The choice of cross-cultural learning strategies and processes provides a best-suited environment for generating unconventional ideas and innovative conceptual schemes. Besides, the sharing of experiences, know-how and skills enables a high-level capability of Human Capital for the digital innovation concept, for exploring new organisation models and, in some cases, for research spin-offs. The interactions among the cross-cultural learning strategies, and the "socialisation" process arising from the partnerships enhance the process of generating new ideas that, according to Garvin, are essential if learning is to take place. Sometimes they are created *ex novo*, through flashes of insight or creativity; other times they come from outside the organisation, or are communicated by knowledgeable insiders (Garvin, 2000).

5.6.2 *The eBMS organisation*

The eBMS strategy has been put into practice through an innovative organisational model. The eBMS organisational structure model and its resources are described below.

The eBMS organisational structure

The eBMS organisational structure is shown in Fig. 5.4.

- *"Talented people"* are the human resources involved in masters, PhD, undergraduate and executive learning programs. Students come from

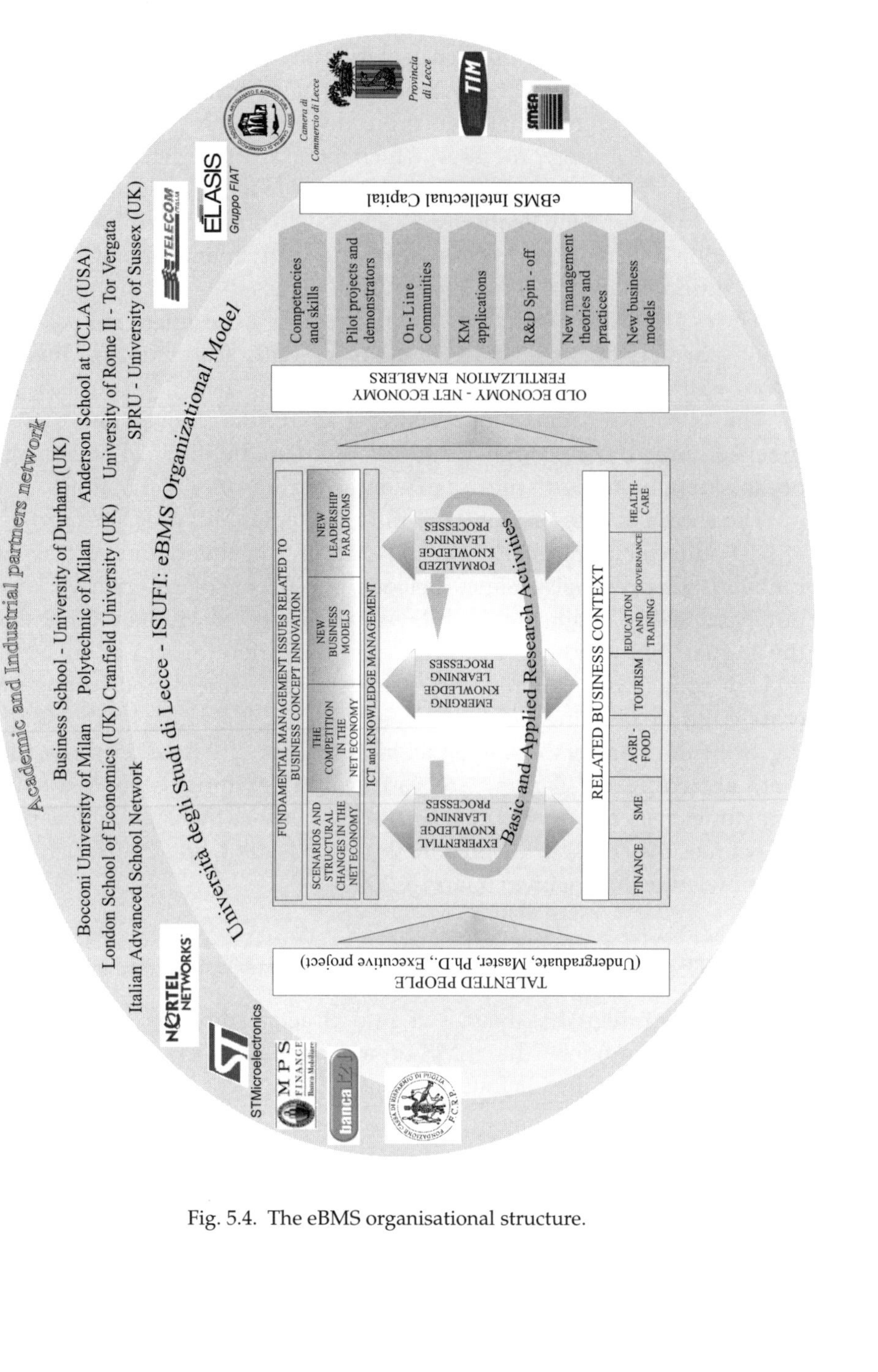

Fig. 5.4. The eBMS organisational structure.

Italy and from other Mediterranean countries. Executives, researchers and managers, mainly involved in the Summer School, come from all over the world;

- "*Academic and industrial partners network*" represents all the organisations currently involved in the experiment;
- "*Fundamental management issues*", related to context innovation, highlight the cross-cultural approach, including organisational, technological and managerial components, concerning:
  - Scenarios and structural changes in the Digital Economy;
  - Competition in the Digital Economy;
  - New organisation models;
  - New leadership paradigms;
  - ICTs and knowledge management.

  These fundamental management issues aim at offering a theoretical framework related to:

  - The emerging dynamics of organisation transformation in the Digital Economy;
  - Some new interpretative approaches of these new emergent dynamics;
  - The shift from traditional innovation management to the digital innovation concept;
  - Digital Innovation Leadership as a new leadership paradigm;
  - Digital Innovation Leadership as Knowledge Leadership;
  - The organisational knowledge creation process as a fundamental process of Knowledge Leadership;
  - The technological tools and solutions to support the Internet-worked organisation;
- "*Related organisation context*" refers to the traditional organisation fields in which the Old Economy — Digital Economy fertilisation process is explored and experienced;
- The eBMS engine is provided by the basic and applied research activities, which boost the creation and exploitation of experiential knowledge (often tacit, related to specific events, experiences, trials, observations), emerging knowledge (explicit knowledge designed to meet the needs of a specific situation) and formalised knowledge (explicit knowledge generalised and structured to educate) (Ives, Torrey, and Gordon, 2000). The basic and applied research activities are strictly interrelated: indeed it is presumed that theory and practice foster each other in a synergic manner. The goal of basic research is to predict the transformation induced by structural changes in the Digital Economy;

the outcomes of basic research are necessary to identify new opportunities coming from such transformations. The outcomes facilitate the development of applied research — such as pilot research projects — aimed at experimenting new organisation cases that are based on the new opportunities which emerged during the basic research activities. Basic research is also focused on the assessment of the real opportunities for SMEs to increase their competitiveness, by adopting pilot project outcomes. In-progress and final assessment of pilot projects can also provide feedback to enrich the value of the knowledge base generated by basic research activities, possibly validating or updating their contents.

Basic and applied research activities are based on the following items:

a) The definition of a Digital Economy framework through the analysis of the structural changes caused by the introduction of ICT/Internet, and the understanding of the socio-economic network dynamics, that provides the tools to govern the transition from the Old to the Digital Economy. The research agenda is:

   - The transition toward the Digital Economy: determining the extent and the speed at which the following processes are taking place : first, vertical disintegration of the organisation, second, fragmentation of the value chain and third, recombination of the agents into networks;
   - The dynamics of network, that takes account of the most recent results arising from complexity theory, evolutionary economics and knowledge management. In particular the research examines the phenomenon of self-organisation, the correlation between self-organisation and network stability and management criteria compatible with self-organisation;

b) The internetworking technologies and the evaluation of their impact as qualifying factors for organisational and managerial transformations. Particularly we will focus on:

   - Analysis of the technological platforms for Internetworking;
   - Convergence of technological platforms;
   - Systems of Enterprise Resource Planning (ERP);

c) The study of new models of organisation and of region development. Moving from the analysis of the impact of Internet-based applications on intra- and inter-organisation relations, we focus on the analysis and development of:

- e-Business models in several economic and social sectors.
- Digital and relational connectivity models (virtual productive districts and Smart Communities) in order to position organisations and regions in the competitive environment of the Digital Economy;

d) The design, development and prototyping of pilot applications, as demonstrators of the benefits arising from the new organisation assets allowed by the Digital Economy. The general framework of the pilot applications includes:

- The setting up of an "e-Organisation" innovative model both for the SMEs (in the agrifood and tourism industries) and for education, finance, government institutions, and healthcare;
- The development of an open platform prototype — the marketplace — equipped with technological tools, protocols and mechanisms, focused on the supply chain relationships and processes, adequate to support the concept of extended virtual organisations, especially in SME clusters;
- The integration of the marketplace platform with wireless Internet devices and with intelligent technologies for tracking physical objects; The marketplace platform, for the applications, is based on the following software tools:
    - "Customer Relationship Management" (CRM);
    - "Supply Chain Management" (SCM);
    - "Enterprise Resource Planning" (ERP);
    - "Business Intelligence" (BI);
- The setting up and diffusion of new management best practices;
- The setting up of methodologies and models for monitoring the evolution of Internet-based technologies, and for evaluating their impact on intra and inter-organisational assets;

- Old Economy — Digital Economy fertilisation enablers that represent the macro-results of the experiment. These enablers are expressed in terms of competencies and skill levels increased, pilot projects and demonstrators designed and rewarded, on-line communities activated, knowledge management applications realised, R&D spin-offs, new management theories and practices investigated, new organisation models designed. All these items contribute to eBMS Intellectual Capital growth.

The eBMS core processes

Core processes are chosen for encouraging entrepreneurial behavior in talented people, stimulating them to consider change as an opportunity, to understand the value of the exploration spirit and of openness to changes. Indeed, core processes are selected in order to increase the performance of Human Capital in leading the Old Economy — Digital Economy fertilisation process. In this perspective, the core processes are conceived so as to induce talented people to:

- Learn to think "outside the box";
- Trust to think and act strategically;
- Learn how to define and acquire knowledge;
- Learn how to learn;
- Understand the knowledge levers for competitive advantage;
- Recognise the organisation as a social structure;
- Understand how to increase value through strategic alliances, economic webs, teamwork processes, new balance between control and empowerment.

These concepts suggest we search for the best relationships existing between organisational and knowledge creation learning processes. In this view, the experiment aims at setting up the most suitable organisational learning process for supporting the knowledge creation processes framed by Nonaka and Takeuchi in terms of "SECI process".

In the eBMS experiment, this theoretical framework is applied expanding on and contextualising the operational model of the "SECI process" developed by Nonaka and Takeuchi (1995). Each step of the SECI process is associated with a different level of knowledge that ranges from experiential knowledge, to emerging knowledge and formalised formalised knowledge (Ives, Torrey, and Gordon, 2000).

The socialisation step aims at developing mainly experiential knowledge, often tacit, related to specific events, experiences, trials, observations, and empirical analyses. This "Experiential knowledge" learning process is developed mainly through workgroups and open forums with trained people, belonging to academic and industrial partners. These events are useful for talented people who wish to share informal insights on specific subjects.

The externalisation step concerns mainly the "Emerging knowledge" learning processes that develop during the setting-up and realisation of pilot projects and specific research programs, aimed at realising new organisation applications that support Old Economy-Digital Economy fertilisation.

Combination and internalisation phases are much more focused on cross-leveling knowledge, which concerns "Formalised knowledge" learning processes, aimed at obtaining three fundamental results:

- Systematising concepts and conceptual schemas, incubated in the experiential and emerging learning processes, into advanced and up-to-date theoretical models;
- Triggering "experiential and emerging knowledge" learning processes, generating interests and motivations for individual and collective experiences;
- Offering a contribution for setting up a theoretical framework useful to systematise further experiential and emerging knowledge.

All four steps are dynamically interdependent, enhancing individual and collective performances within the eBMS.

All three processes provide value and need to be captured and managed within the eBMS organisation, with different methodologies. Indeed, they are all addressed within the eBMS, through forums or discussions for experiential knowledge, libraries and project deliverables for emerging knowledge, that are maintained and monitored by professional staff, an methodologies of formal knowledge, that are created by dedicated staff and then tested and revised through the work of BMS users.

The eBMS resources

The strategic resources (core competencies and physical/informational resources) have been selected according to the strategy choices and organisational assets.

Core competencies

The fundamental characteristic of the experiment is a continuous search for benefits and values deriving from network effects. In this perspective, core competencies are made up both of local specialised people, and of outstanding people coming from academic and industrial partners.

Core competencies are specialised for each kind of knowledge:

- As regards the learning processes related to experiential knowledge, eBMS has a staff, made up of professionals, that guides and supports students in their knowledge-sharing activities, through help-desk tools, newsletters, books and other means; this is the staff who:
  - Organise lectures;
  - Set up laboratories and select smart lecturers;
  - Create spaces (virtual and otherwise) to enhance student-student and student-professor or tutor interactions;
  - Build and develop the eBMS community through initiatives that involve community members in an enjoyable learning environment;
- As for the learning processes related to the knowledge emerging in projects, eBMS competencies include:
  - People keen on setting up relationships with business partners, and developing the knowledge, insights, tools and techniques for management practice, i.e. for carrying on rigorous and innovative research that improves management theory and practice, through the following guiding principles:
    - Attracting, developing and retaining outstanding faculty, testimonials, business partners and staff who produce excellence in research and education;
    - Building linkages that enhance and exploit the eBMS' distinctive intellectual excellence;
    - Developing a working environment based on mutual respect, high ethical standards and productive collaboration;
  - Experts in attracting business partners: eBMS has strong ties with industry and develops common programs. This is useful for the eBMS itself (business partners enable us to touch on real organisation problems, and are strategic testimonials in educational activities) and for its business partners too (who can find innovative solutions for their businesses). Experts are always looking for new alliances and better ways to establish mutually advantageous relationships with businesses around the world;
- As regards learning processes related to formal knowledge, all eBMS actors are given time away from their formal tasks to focus on deliverables in terms of formal knowledge which is then used and refined by students, teachers, staff, alumni, visiting professors, visiting scholars, academic and organisation partners.

Physical and informational resources

The choice of the physical and informational resources is related to the following goals:

- Continuous scanning of organisation transformations on a global scale, in order to see the emerging patterns, best practices, paradigm shifts and opportunities arising from the competitive landscape of the Digital Economy;
- Creating opportunities, at an international scale, for synergistic thinking and co-operative research aimed at re-thinking, re-directing, re-organising and re-building theoretical and practical organisation frameworks;
- Supporting the evolution towards organisational processes based on a "virtualness strategy".

These goals trigger the realisation of the following strategic assets:

a) The empirical laboratory specialising in monitoring the emerging phenomena of the Digital Economy;
b) The Advanced International Summer School on "e-Business and complexity: new managerial practices";
c) The "Virtual eBMS" infrastructure enabling information resource management, knowledge bases and collaborative technologies.

Below is a brief description of each asset.

a) The laboratory on "Emerging phenomena of the Digital Economy" is conceived as a permanent observatory of the emerging organisation innovations in the Digital Economy landscape. The methodology is characterised by Internet-based research, for gathering primary and secondary data, useful for answering the research questions of the innovation agenda.

   More particularly, research questions concern the impact of the Internet-driven innovations on the organisation, in terms of:

   - Core strategy: organisation mission, product/market scope, basis for differentiation;
   - Strategic resources: core competencies, strategic assets, core processes;
   - Customer interface: fulfillment support, information and insight, relationship dynamics, pricing structure;
   - Value network: suppliers, partners, coalitions;

- Configuration of activities;
- Customer benefits;
- Organisation boundaries;
- Performances obtained.

The empirical laboratory includes case studies that are emerging in the manufacturing sectors ("specialised suppliers", "science based" organisations, "scale intensive" organisations, "supplier dominated" organisations) and in traditional and innovative services.

b) The Advanced International Summer School on "e-Business and complexity: new managerial practices".

The Summer School aims at providing a multidisciplinary forum for academics, managers and consultants to discuss the managerial implications of the Digital Economy in an international setting. The style is informal, the range of disciplines and sectors intentionally broad, the venues chosen for fostering communities around ideas and projects. Therefore, the mission of the Summer School is a periodic up-dating, at international level, of the theories, techniques and policies concerning the new management practices in the "knowledge-based" Digital Economy. More specifically, the School's objective is to provide an international workshop for researchers, scholars and top managers, engaged in different research fields, in order to debate, from different points of view, topics concerning the emerging patterns of Digital Economy and to discuss new management practices related to these patterns.

The workshops organised at the Advanced International Summer School are structured in three main activities in which academics and professionals are involved:

- Academic and organisation-led presentations;
- Group-based discussions: participants debate key issues introduced in the plenary presentations;
- Panel discussions: speakers and group leaders summarise the day's discussions.

Top managers from organisations in highly innovative sectors together with consultants and leading academics debate and analyse the contribution that new theories can offer strategy and management, in terms of approach, frameworks and techniques.

The convergence of academics, consultants and practitioners also constitutes a rare chance to build up a community to pioneer the applications of complex systems thinking in industry. Beyond the objectives of learning and forming networks, the School aims at:

- Gaining an understanding of the gap between current management practices and new ideas emerging from complexity thinking;
- Setting up a research agenda concerning the implications of complexity thinking in the Digital Economy focused on the following topics:
  - The strategic importance of learning and knowledge in the Digital Economy;
  - Developing strategy and leadership approaches for the Digital Economy: methods of understanding the dynamics of complex adaptive systems using agent models. New approaches for strategising under uncertainty and facilitating self-organisation;
  - Defining new organisation models.

c) The "Virtual eBMS" infrastructure is conceived as the key lever of the eBMS organisational knowledge creation processes.

The infrastructure supports the development of an eBMS Virtual Learning Community that is a geographically distributed collaborative environment that allows all the partners' activities to be integrated. In such a Virtual Learning Community, we want to produce an interactive and strongly stimulating environment in which we could integrate lectures, courses, texts, graphics, with simulations, case studies and virtual classrooms. In such a Community, the cognitive process is not oriented to the stimulus-answer model, but to a model of non-sequential knowledge. The opportunities offered by the eBMS Community, constituted by actors or stakeholders coming from different industries and academies, facilitates the circulation of knowledge and allows value to be added to the different activities. In this way, the Virtual Learning Community is a concrete and realistic tool for creating a favorable environment for the creation of new knowledge through both practical and theoretical cases. A first categorisation of the community members can be done by referring to the actual actors in the eBMS community, who are: partners, ex students, lecturers (invited for talks), testimonials, researchers (involved in research projects), students, and staff.

In this perspective, the Virtual eBMS infrastructure is conceived as the space of interaction among the members of the eBMS. Some typical interactions that may take place in this space are:

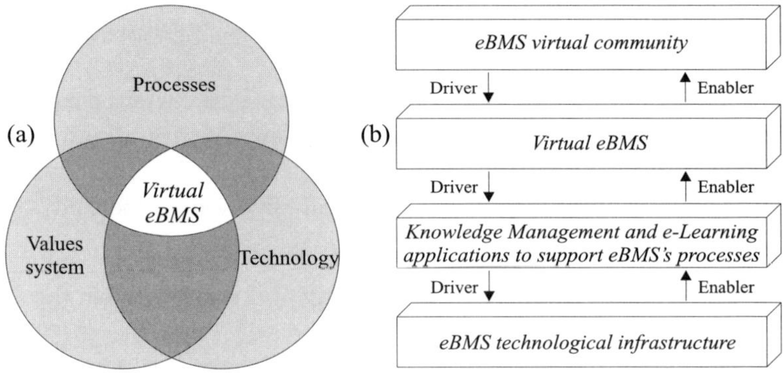

Fig. 5.5. The interacting components and the dynamic model of the virtual eBMS (Elia and Corallo, 2001).

- Information exchanges;
- Resource sharing;
- Learning events;
- Collaborative work.

Therefore, the Virtual eBMS infrastructure represents a specific strategic project of the eBMS experiment. The general schema of this project (Fig. 5.5) shows the interacting components and the dynamic model of the virtual eBMS:

This schema describes a dynamic framework in which each level is related to the others according to a driver/enabler relationship. In this framework, the value system and the processes described in the eBMS model drive the features of the technological infrastructure and of the applications. Indeed, the "eBMS virtual community" expression summarises the set of people, value system, and processes operating in the eBMS (Elia and Corallo, 2001).

5.6.3 *The eBMS environment*

As previously highlighted, strong and varied partnership is a strategic component of the basis for differentiating the experiment. In order to gain benefits and value from network effects, a wide partnership program has been developed. This program includes relationships with:

1. Suppliers, to develop customised technological solutions to support our advanced learning processes and research activities;

2. Partners, selected to integrate the eBMS core competencies and strategic assets, particularly for the experiential and emerging knowledge processes. The eBMS staff faculty includes outstanding people from several Universities, research centers and organisations at a global level.

The value network related to the organisation community is also enriched by:

- Advisory boards with leading organisation and other personalities, which meet regularly and give sponsorships;
- Numerous organisations recruiting from the eBMS;
- Scholarship programs by organisations;
- Awards by organisations for outstanding students.

Besides, the eBMS experiment grasps the benefits coming from its network, made up of top-flight academic partners.

As regards the eBMS's customers, the nature of the experiment has blurred the boundaries between employees and customers. In this view, we adopted the term "user" rather than "customer". Actually, the users of eBMS are the students, the researchers, the staff faculty members, while the partners are more generally all the actors receiving value from the eBMS. A very distinctive characteristic of the eBMS is that the users form a community of practice, with no geographical boundaries, involved in creating Organisation Innovation. The main motivation for the community's growth and development is the integration of experiential knowledge, emerging knowledge and formalised knowledge process.

In this perspective, the interacting components and the dynamic model of the virtual eBMS, described in the previous paragraph, enable and improve the level of user participation to the community members.

5.6.4 *The performance of the eBMS experiment*

The search for appropriate parameters to measure the performance of the experiment, in terms of value created, falls within the general issues concerning "Intellectual Capital assessment" (Sveiby, 2000). The "Balanced Scorecard" concept suggested by Kaplan and Norton has been successful in highlighting the importance of non-financial measures in assessing the long-term health of an organisation (Kaplan and Norton, 1996). However, according to Laurence Lock Lee, measures like customer loyalty, staff satisfaction level, R&D expenditures and revenue per employee, all indicate results without giving any insight into the processes that have led to the results being achieved or the cause/effect relationship (Lock Lee, 2000).

The distinctive features and results of the dynamics of the eBMS experiment are mainly related to its processes, such as experiential knowledge learning, emerging knowledge learning, formalised knowledge learning. The learning processes, enabled in different ways by interactions among individuals, teams, organisations, partners, need to be evaluated according to a more appropriate methodology than the Balanced Scorecard. As stated by Lock Lee, a best-suited evaluation model for knowledge sharing, management and creation process is "in process" metrics.

Indeed, Total Quality Management (TQM) practitioners have demonstrated that superior quality may be achieved without the need for extensive inspection checks, by keeping processes "in control". Therefore, in order to evaluate the macro-results of the eBMS experiment, "in process" metrics suggest choosing the following as the main indicators of the eBMS experiment:

- The eBMS-related growth of competencies and skill levels achieved;
- Pilot projects conceived and rewarded;
- On-line communities designed and implemented;
- Knowledge management applications developed;
- New management theoretical approaches and practices explored;
- New organisation models envisioned;
- R&D spin-offs generated.

Here I provide some insights about the preliminary results of the experiment that started in November 1999.

Competencies and skill levels

In the three years of the experiment, almost 60 young talented people have been selected from approximately 180 applicants and have gone through the eBMS learning processes. About 30% came from non-EU countries in the Mediterranean Area. The main competencies developed and exploited concern the ability to conceive and design e-Organisation models, to be applied to various domains of the Old Economy. Almost 80% of all the young talented people participating in the eBMS learning process are now employed at high levels in prestigious manufacturing and service organisations.

e-Business pilot projects

Some pilot projects have already been started, launched and rewarded during the eBMS life cycle:

- FoodNetXchange;
- VISPO;
- CRM for e-Finance.

The FoodNetXchange project is an e-Marketplace for organisation innovation in the agrifood industry. A consortium of 18 partners based at the eBMS was finally established on July 2001 to lead the FoodNetXchange project. The consortium was constituted from organisation partners operating at every level of the agrifood chain (R&D Labs, Universities, agrifood producers, distributors, services providers, logistics players, etc.). Partners have defined their strategic tasks and roles within the FoodNetXchange e-Business Community.

The VISPO project aims at setting up a prototype of an Application Service Provider (ASP), able to offer an integrated set of customised services to the SMEs of an industrial district. The identification of services was based on a preliminary analysis of the industrial districts needs, in terms of inter-organisation integration, economies of scale, support to the brand of the district. The VISPO project is a joint research project involving eBMS, ENEA, TELECOMLAB, Consiel and the Polytechnic of Milan.

The CRM project for e-Finance aims at designing a CRM application for the financial services industry, in order to broaden the retail channels. This is a joint pilot project involving MPS-Finance and eBMS.

On-line communities models and architectures

Two "on-line communities" have been conceived and designed:

- The on-line Community in the TIM Fellowship Programme;
- The Knowledge Hub Enabling Regional Innovation Leadership.

The "TIM Fellowship Programme" community aims to build a strong synergy between TIM (Telecom Italia Mobile), people, and academia. The main services available in the community are: forums, e-Learning, e-Library, oriented toward the creation of a virtual incubator of organisation ideas.

The Knowledge Hub Enabling Regional Innovation Leadership aims to establish an on-line community project to foster local communities' capabilities to dynamically recognise their competitiveness status and then to continuously agree on new and/or updated actions. The aim of the community project is to promote cluster-based development, foster interactive learning and innovation, assist in developing local policy institutions with a better performance in diminishing lock-in and institutionalised thinking, understand market-based and cooperatively-based knowledge governance mechanism under the conditions of the Digital Economy.

New organisation models

Among the basic research activities carried on at eBMS, three main track themes have been focused on:

- Digital Economy interpretative approaches;
- The Hypernet revolution;
- The impact of Internet technologies on SME relationships and competitiveness.

Digital Economy interpretative approaches aimed at gaining a deeper understanding of the phenomena related to Digital Economy in order to study and envision new development paths and processes for organisations in the new arena.

The Hypernet revolution research program aims at investigating the rapidly changing nature of the Internet and its possible evolutions, in particular its impact on organisation and social activities. The purpose of this program is to carry out a preliminary exploratory research on how emergent characteristics of the Hypernet can boost pervasive and non-linear innovations in the organisation environment in order to extract some useful guidelines for the creation of Digital Innovation Leadership and for further developments of the research itself.

The research agenda on the impact of Internet technologies on SME relationships and competitiveness intended to explore the impact of Internet technologies on commercial relationships involving small firms in specific industries. The aim is to figure out possible changes to transaction structure and opportunities for small firms to increase their competitiveness.

5.7 Conclusions

Digital innovation is boosting the shift from hierarchical capitalism, with large and integrated corporate hierarchies, to flexible capitalism, with more inter-firm alliances, single firm heterarchies and corporate networks (Dunning, 1997). Since these networked organisations exchange a large amount of information and knowledge, managing networks in the Digital Economy requires a new type of leadership. More specifically leadership becomes a "collective capacity", and more specifically a capacity to manage knowledge, that has become mobile and global in the new networked assets.

As a consequence, the need for Knowledge Leadership is emerging, and this has become the main challenge for Business Schools. To face this challenge, it becomes necessary to introduce radical innovations in the knowledge bases and in the learning strategies of Business Schools, to support the creation of Knowledge Leadership.

This is the view of "e-Business Management School" (eBMS), an experiment that we are currently undertaking at the ISUFI-University of Lecce (Italy).

The eBMS aims at realising a Knowledge Leadership Incubator, where talented people can learn how to learn together, enhancing their collective intelligence, competence and capacity to imagine research questions and pilot projects oriented to generate innovation in the Digital Economy environment.

The above overview of the preliminary results highlights the high performance of the eBMS experiment. The success of the experiment presented in this chapter is particularly due to an environment where value is the result of a network of academic and industrial partners.

Among the other determinants, we point out:

- The learning strategies practised in the eBMS core processes;
- The knowledge management technological infrastructure, which has enabled the eBMS to act as a Virtual Community of Practice.

References

Amit R. and Schoemaker P. (1993) Strategic assets and organisational rent. *Strategic Management Journal* **14**, 33–46.

Amit R. and Zott C. (2001) Value creation in e-business. *Strategic Management Journal* **22**, 493–520.

Armstrong A. and Hagel J. (1996) The real value of on-line communities. *Harvard Business Review*, May–June, 134–141.

Barney J.B. (1991) Firm resources and sustained competitive advantage. *Journal of Management* **17**, 99–120.

Bessant J. and Tsekouras G. (2001) Developing learning networks. *AI & Society* **15**, 82–98.

Davis S. and Botkin J. (1994) The coming of knowledge-based business. *Harvard Business Review*, September–October, 165–170.

Dunning J.H. (1997) A business analytic approach to governments and globalisation, in Dunning J.H. (ed.) *Governments, Globalisation and International Business*. Oxford University Press, London, 114–131.

Eisenhardt K.M. and Martin J.A. (2000) Dynamic capabilities: What are they? *Strategic Management Journal* **21**, 1105–1121.

Elia V. and Corallo A. (2001) Towards the virtual eBMS, in Romano A., Elia V. and Passiante G. (eds) *Creating Business Innovation Leadership. An Ongoing Experiment*: *The e-Business Management School at ISUFI*. Edizioni Scientifiche Italiane, Naples.

Elia V. and Passiante G. (2001) Setting the framework for business innovation leadership, in Romano A., Elia V. and Passiante G. (eds) *Creating Business Innovation Leadership. An Ongoing Experiment*: *The e–Business Management School at ISUFI*. Edizioni Scientifiche Italiane, Naples.

Evans P. and Wurster T.S. (1997) Strategy and the new economics of information. *Harvard Business Review*, September–October.

Gulati R., Nohria N. and Zaheer A. (2000) Strategic net-works. *Strategic Management Journal — Special Issue* **21**, 203–215.

Haeckel S.H. (1999) *Adaptive Enterprise*: *Creating and Leading Sense and Respond Organisations*. Harvard Business School Press, Boston.

Hagel J. and Rayport J.F. (1997) The coming battle for customer information. *Harvard Business Review*, January–February, 53–65.

Handy C. (1995) Trust and the virtual organisation. *Harvard Business Review*, May–June, 40–50.

Iansiti M. and MacCormack A. (1997) Developing products on internet time. *Harward Business Review*, September–October, 108–117.

Ives W., Torrey B. and Gordon G. (2000) Knowledge sharing is a human behavior, in Morey D., Maybury M. and Thuraisingham B. (eds) *Knowledge Management*: *Classic and Contemporary Works*. The MIT Press, Cambridge.

Janszen, F.H.A. (1997) Innovation and the materials revolution. *Technovation* **17**, 549–556, 594.

Kaplan R.S. and Norton D.J. (1996) *The Balanced Scorecard: Translating Strategy into Action*. Harvard Business School Press, Boston, Massachusetts.

Lock Lee L. (2000) Knowledge sharing metrics for large organisations, in Morey D., Maybury M. and Thuraisingham B. (eds) *Knowledge Management: Classic and Contemporary Works*. The MIT Press, Cambridge.

Malone T.M. and Laubacher, R.J. (1998) The dawn of the e-lance economy. *Harvard Business Review*, September–October.

McKenna R. (1995) Real-time marketing. *Harvard Business Review*, July–August, 87–95.

Moore J.F. (1993) Predators and prey: A new ecology of competition. *Harvard Business Review*, May–June, 75–86.

Nishiguchi T. (2001) Coevolution of interorganisational relations, in Nonaka I. and Nishiguchi T. (eds) *Knowledge Emergence*. Oxford University Press, London.

Nolan R. and Haeckel S.H. (1993) Managing by wire. *Harward Business Review*, September–October.

Nonaka I. and Nishiguchi T. (2001) Social technical and evolutionary dimensions of knowledge creation, in Nonaka I. and Nishiguchi T. (eds) *Knowledge Emergence*. Oxford University Press, London.

Nonaka I., Toyama R. and Konno N. (2001) SECI, Ba and leadership: A unified model of dynamic knowledge creation, in Nonaka I. and Teece D. (eds) *Managing Industrial Knowledge*. Sage Publications, London, 13–43.

Penrose E.T. (1959) *The Theory of Growth of the Firm*. Basil Blackwell, London.

Peteraf M.A. (1993) The cornerstones of competitive advantage: A resource-based view. *Strategic Management Journal*, **14**, 179–191.

Polanyi M. (1966) *The Tacit Dimension*. Routledge & Kegan Paul, London.

Porter M.E. (1985) *Competitive Advantage: Creating and Sustaining Superior Performance*. The Free Press, New York.

Prahalad C.K. and Hamel G. (1990) The core competence of the corporation. *Harvard Business Review*, May–June, 79–91.

Prusak L. and Cohen D. (2001) How to invest in social capital. *Harvard Business Review*, June, 86–93.

Reicheld F.F. and Schefter P. (2000) E-loyalty. *Harvard Business Review*, July–August.

Romano A. (2001) Introduction to creating business innovation leadership: Exploring new theoretical and practical approaches, in Romano A., Elia V. and Passiante G. (eds) *Creating Business Innovation Leadership*.

An Ongoing Experiment: The e-Business Management School at ISUFI. Edizioni Scientifiche Italiane, Naples.

Romano A., Elia V. and Passiante G. (eds) (2001) *Creating Business Innovation Leadership. An Ongoing Experiment: The e-Business Management School at ISUFI*. Edizioni Scientifiche Italiane, Naples.

Romano A., Passiante G. and Elia V. (2001) New sources of clustering in the digital economy. *Journal of Small Business and Enterprise Development* **8**(1), 19–27.

Senge P. (2000) Reflection on a leaders new work: Building learning organisations, in Morey D., Maybury M. and Thuraisingham B. (eds) *Knowledge Management: Classic and Contemporary Works*. The MIT Press, Cambridge, 53–60.

Schumpeter J.A. (1934) *The Theory of Economic Development: An Inquiry into Profits, Capital, Credit, Interest, and the Business Cycle*. Harvard University Press, Cambridge, Massachusetts.

Seeman P. *et al.* (2000) Building intangible assets: A strategic framework for investing in intellectual capital, in Morey D., Maybury M. and Thuraisingham B. (eds) *Knowledge Management: Classic and Contemporary Works*. The MIT Press, Cambridge, 85–98.

Sveiby K.E. (2000) Measuring intangibles and intellectual capital, in Morey D., Maybury M. and Thuraisingham B. (eds) *Knowledge Management: Classic and Contemporary Works*. The MIT Press, Cambridge, 53–60.

Tapscott D., Ticoll D. and Lowy A. (2000) *Digital Capital*. Nicholas Brealey Publishing, London.

Teece D.J., Pisano G. and Shuen A. (1997) Dynamic capabilities and strategic management. *Strategic Management Journal* **18**, 509–533.

Upton D.M. and McAfee A. (1996) The real virtual factory. *Harvard Business Review*, July–August, 123–133.

Wernerfelt T.B. (1984) A resource-based view of the firm. *Strategic Management Journal* **5**, 171–180.

Williamson O.E. (1975) *Markets and Hierarchies, Analysis and Antitrust Implications: A Study in the Economics of Internal Organisations*. Free Press, New York.

Glossary

Agent-based models. Simulation models that are the favourite tools of those who practice the interdisciplinary Sciences of Complexity. Agent-based modellers typically instantiate some software agent with some limited form of intelligence, also known as a constraint, and unleashes them in an environment designed to accommodate interactions among many agents. This environment is often modelled on the real world so that interactions are reflective of naturally occurring contacts with each other.

AgriFood sector. Firms involved in the agriculture, food and biotechnology industries.

ASP (Active Server Page). An HTML page that includes one or more scripts (small embedded programs) that are processed on a Microsoft Web server before the page is sent to the user.

ASP (Application Service Provider). A company that offers individuals or enterprises access over the Internet to applications and related services that would otherwise have to be located in their own personal or enterprise computers.

Ba. "Shared context", where knowledge is created, shared and diffused.

Balanced Scorecards. Method of measuring and managing business performance giving a balanced view of financial and operational perspectives to accelerate the management process.

Business Concept Innovation. Theoretical framework for innovation management, that links innovation creation to all the components of the business. The framework merges into an overall model the different views of business that typically come from different disciplines: marketing, strategic management, organisation.

Business Intelligence (BI). Systems that collectively hold explicated formal and informal knowledge and the rules associated with them, for accumulating, refining, managing, validating, maintaining, annotating (adding context) and distributing the contents.

Community of Practice (CoP). Special type of informal network that emerges from a desire to work more effectively or to understand work more deeply among members of a particular specialty or work group. At the simplest level, CoPs are small groups of people who have worked together over a period of time and through extensive communication have developed a common sense of purpose and a desire to share work-related knowledge and experience.

Community Network value. Sum of all the individuals' value of a network.

Competitiveness. Characteristics of firms or industries believed to lead to a strategic market advantage. For example, a firm might be considered to be competitive if it is the low cost leader in its industry, if it has focused on a market niche, or if it has a superior technology, or some other sustainable advantage (Porter 1980, 1985). However, if such a firm does not capitalise on its advantage to achieve high market share and profitability, it is not competitive in the long run. The conditions that may yield competitive advantage for a community, county group, region or nation are more ambiguous and potentially varied (Porter 1990). The outcomes must be viewed in terms of long term, stable improvements in the quality of life of the residents and the ability to support and nurture enterprises capable of bringing resources from the environment into the community.

Complex adaptive systems (CAS). Systems made up of large numbers of interactive adaptive elements that are diverse in form and capability. The complexity results from the adaptive behavior of the agents.

Complex systems. Structures and processes that involve non-linearity. This includes both sets of many interacting objects as well as non-linear dynamical systems.

Complexity theory. Scientific theory related to self-organisation.

Connectivity. Access to the global digital network that includes satellites, cable TV, new wireless systems, mobile telephones and the Internet.

Creative destruction. Schumpeterian definition of innovation processes that are destroying old regimes and creating new, more exciting assets, through less predictable scenarios and opportunities.

Customer Relationship Management (CRM). Combination of business processes and technology that seeks to understand a company's customers from multiple perspectives in order to competitively differentiate a company's products and services. The focus of CRM is to improve levels of customer satisfaction, boost their loyalty, and increase revenues from existing customers.

Digital connectivity. A total blur in digital form of formerly discrete messages: text, sound, voice, phone, cable, pictures, movies, TV, radio etc. These messages are obviously capable of transmission and receipt by identical vehicles.

Digital Economy. An economy based on the digitisation of information and the respective information and communication infrastructure. This new form of an economy implies not only technological, but also and especially process-related and structural challenges and potentials. The way in which economic values are created will change fundamentally in the Digital Economy and thus transform the general structure of economies and societies.

Digital innovation. Innovation enabled by digital technologies.

Economic development. Qualitative change and restructuring in a country's economy in connection with technological and social progress. The main indicator of economic development is increasing GNP per capita (or GDP per capita), reflecting an increase in the economic productivity and average material well-being of a country's population. Economic development is closely linked with economic growth.

Economic growth. Quantitative change or expansion in a country's economy. Economic growth is conventionally measured as the percentage increase in gross domestic product (GDP) or gross national product (GNP) during one year. Economic growth comes in two forms: an economy can either grow "extensively" by using more resources (such as physical, human, or natural capital) or "intensively" by using the same amount of resources more efficiently (productively). When economic growth is achieved by using more labour, it does not result in per capita income growth. But when economic growth is achieved through more productive use of all resources, including labour, it results in higher per capita income and improvement in people's average standard of living. Intensive economic growth requires economic development.

e-Business. Internet-facilitated execution, coordination and management of business processes and activities. Electronic commerce is a subset of e-Business.

e-Finance. A growing and diversifying area of e-Business, that, from net-banking, e-Trade to other financial businesses, covers almost all aspects of personal finance such as insurance, mutual funds and personal loans.

Emerging knowledge. Both tacit and explicit knowledge, formal but not generalised. It is mainly related to newsletters, proposals and deliverables, and produces structural thinking.

Enterprise Resource Planning (ERP). Combination of business processes and technology that attempt to integrate all departments and functions across a company onto a single system that can serve all those different departments' particular needs, so that the various departments can more easily share information and communicate with each other.

Explicit knowledge. Anything that can be documented, archived and codified, often with the help of Information Technologies

Externalities. Any occurrence or activity that lies outside the range of control of the individual firm, but that then has definite effects on the firm's internal production function.

Experiential knowledge. Knowledge, often tacit, related to specific events, experiences, trials, observations, empirical analyses. Generally it produces insights and understandings.

Evolutionary economics. Theory that explains economic behaviors in terms of mechanisms of preservation and transmission, mechanisms of variety-creation and mechanisms of selection.

Forecasting models. Models based on the old cause-effect beliefs.

Foreign Direct Investment (FDI). It reflects the objective of obtaining a lasting interest by a resident entity in one economy (direct investor) in an entity resident in an economy other than that of the investor (direct investment enterprise).

Fourfold spatial hierarchy. A type of economic and political relationship. It is made up of:

- The global level constituted by different networks of economic activity;
- The pluri-national level represented by a group of multination blocs, such as the EU, NAFTA, etc.;
- The national level focused on the classical sovereign state;
- The regional level.

Formalised knowledge. Explicit knowledge, generalised and structured to educate. It is related to methods, formal training and mission statements.

General Purpose Technology (GPT). Technology that can be used in many downstream applications because it provides a generic function, such as the developments in communication technologies.

Geographical cluster. Geographic concentrations of interconnected companies, specialised suppliers, service providers, firms in related industries and associated institutions (universities, standard agencies and trade associations) in particular fields that compete but also cooperate.

Geographical proximity. In cluster theory, the co-location of firms in a region.

Global Economy. New competitive environment in which national markets are fused transnationally, rather than linked across borders through flows of trade and investment as in the previous International Economy.

Holistic conceptual framework. Framework in which the whole shows behaviour which cannot be deduced merely by aggregating that of its constituent parts and in which the parts themselves cannot be individually understood separately from the relationships they maintain with each other to make up the whole.

Human Capital. Knowledge, skills and experiences owned by individual employees.

Hypernet Revolution. Business Innovations in the Mobile Economy.

ICT (Information and Communications Technology). It refers to information technology plus telecommunications equipment and telecommunications services.

Increasing returns. Competitive situation where the rich get richer and the poor get poorer.

Industrial district. Groups of SMEs clustered together in specific regions, that seem to be able to grow rapidly, develop niches in export markets and offer new employment opportunities.

Innovation. The search for and the discovery, experimentation, development, imitation and adoption of new products, new processes and new organisational set ups. Transformation of an idea into a marketable product or service, a new or improved manufacturing or distribution process, or a new method of social service.

Institutional economics. Theory that explains the pathway to economic growth and development through the character of established institutions and their ability to adjust to change in the most cost-effective way.

Institutions. A very broad definition meaning:

- Formal organisations (firms, trade unions, universities, state agencies, etc.);
- Patterns of behaviour that are collectively shared (routines, social conventions, ethical codes, etc.);
- Negative norms and constraints (formal laws, moral prescriptions, etc.).

Intangible asset. A claim to future benefits that does not have a physical or financial (a stock or a bond) embodiment. Intangible assets include patents, brands, or unique organisational structures (for example, an Internet-based supply chain) that generate cost savings.

Intellectual Capital. Intangible knowledge assets, including human capital, structural capital and social capital.

Internetworked Enterprise. Enterprise internally connected via Intranet, with suppliers and customers via business-to-business networks and with other organisations, business homes and consumers via public Internet.

Knowledge. Capability for effective action.

Knowledge assets. Inputs, outputs and moderators of the knowledge creation process.

Knowledge base. Large body of "declarative" information used by another program that applies it in the right order to solve a problem (inference).

Knowledge Hub (KH). Knowledge management system that enables innovation at local level, promoting and linking endogenous-exogenous sources of knowledge.

Knowledge Leadership. A collective ability for effective action, embracing both the sharing of what is known and innovation.
"Knowledge Leadership" Incubator. Learning environment that enables the creation of Human Capital, able to generate Structural and Social Capital.

Knowledge Management (KM). Process of managing organisational knowledge for creating business value, sustaining competitive advantages, maximising business growth and value.

Knowledge Management System. Skills and technologies enabling the process of managing organisational knowledge to create business value, sustaining competitive advantages, maximising business growth and value.

Knowledge map. Tool for providing orientation and for supporting the discovery of available knowledge. Existing knowledge domains could be visualised in an easy-to-understand manner.

Learning. The process whereby human beings build knowledge, and enhance their capacity for effective action.

Learning Organisation. Organisation skilled in creating, acquiring, and transferring knowledge, and in modifying its behaviour to reflect new knowledge and insight.

Learning patterns. Learning contents that emerge from the integration of the socio-cultural heritage and educational background of applicants with the learning processes to develop experiential, emerging and formalised knowledge.

Learning region. Region that functions as collector and repository of knowledge and ideas, and provides an underlying environment or infrastructure that facilitates the flow of knowledge, ideas and learning.

Metcalfe's law. The value of the network increases with the square of the number of its participants.

Middleware. Layer of software between the network and the applications. This software provides services such as identification, authentication, authorisation, directories, and security.

National System of Innovation (NSI). A broader definition of NSI includes all parts and aspects of the economic structure and the institutional set-up affecting learning and involved in searching and exploring.

Network Economy. Complex web of transactions based on alliances, linking various parts of an organisation with others.

NUTS. Nomenclature of territorial units for statistics. NUTS nomenclature subdivides the territory of the European Union into 77 at NUTS 1, 206 at NUTS 2 and 1031 at NUTS 3 levels.

Old Economy — Net Economy fertilisation. Process generating business transformations, through the application of Internetworking technologies to the Old Economy.

Ontology. Set of concepts (e.g. entities, attributes) and their relationships that allow a shared understanding of a certain domain of discourse: ontologies provide a shared and common understanding of a domain that can be communicated between people and application systems.

Organisation. A selection of modes of partitioning the system and of modes of linking the parts (Mintsberg, 1979).

Region-enterprise metaphor. Metaphor meaning that, with regard to learning and innovation processes, regional innovation networks behave like Internet-worked Enterprises.

Regional Innovation Leadership (RIL). The ability of a regional community to initiate and sustain significant change in order to work effectively with the forces that shape change.

Regional innovation network. It is a network of regional actors in which the functional boundaries within firms are porous, as are the boundaries between firms, as well as between firms and local institutions such as trade associations, universities, etc. In these networks, firms compete but also learn from each other through informal communication and collaborative processes.

Regional learning community. A community in which individuals and groups continuously expand their ability to achieve the results they desire; in which new and expansive schemes of thought are nourished; in which individuals continuously improve their competencies for "learning together".

Sales & Field Force Automation. Category that includes all the applications that allow sales and technical personnel to stay in constant touch with their head office, access corporate and client information while on the road and submit sales orders without returning to the office.

Diffused sub-categories of these applications are Mobile job dispatch and Field Audit and Inspections.

SECI process. The process of knowledge creation via conversion from tacit to explicit knowledge.

Self-organisation. The evolution of a system into an organised form in the absence of external constraints. The essence of self-organisation is that a system structure often appears without explicit pressure or involvement from outside the system. In other words, the constraints on form (i.e. organisation) are internal to the system, resulting from the interactions among its components and usually independent of the physical nature of those components.

Small and Medium Enterprises (SMEs). According to the EU, a SME is an independent firm with less than 250 employees whose annual turn-over is less than 20m ECU and/or whose balance sheet is less than 10m ECU and in whose capital structure a large firm does not hold more than 25% participation.

Smart Community. A geographical area ranging in size from a neighbourhood to a multi-county region whose citizens, organisations and government institutions "use" ICT to strategically position that region for economic, social and technological development.

Social Capital. Effective networks of relationships with high levels of trust. It includes the ability of groups to collaborate and work together.

Strengths, Weaknesses, Opportunities, and Threats (SWOT) analysis. Examination of an organisation/region's internal strengths and weaknesses, and its environments, opportunities, and threats. It is a general tool designed to be used in the preliminary stages of decision-making and as a precursor to strategic planning in various kinds of applications.

Structural Capital. Rule-based knowledge, embedded both in the organisation's work processes and systems, and in its intellectual property, recognised by patents and copyrights.

Tacit knowledge. Physicist and chemist Michael Polanyi created the concept "tacit knowledge" in the 1950s. Tacit knowledge is difficult to define, but according to one definition it includes all the genetic, bodily, intuitive, mythical, archetypical and experiential knowledge the human being has, even though it cannot be expressed by means of verbal concepts. Tacit knowledge is present in the human being as a whole: it includes manual skills, knowledge of the skin and of thoughts. Tacit knowledge present in the individual guides his or her choices in the information flow. Based on this tacit knowledge the human being can pass by enormous amounts of useless information without ever having to react to it.

Techno-economic paradigm. It is the result of changes in technology systems whose effects have a major influence on the behaviour of the entire economy.

Telematics. The application of information and communications technologies and services, usually in direct combination.

Telemetry. Use of telecommunications to receive or send information to a device.

Total Quality Management (TQM). A process for managing quality: it is a culture of continuous improvement for improving customer satisfaction.

Value network. A web of relationships that generates economic value and other benefits through complex dynamic exchanges between two or more individuals, groups or organisations.

Virtual Cluster. A cluster of organisations, where each member adds one or more distinct aspects of product/service value to the value of the network, by exchanging digital knowledge with other members.

Virtual Learning Community. Collaborative environment, geographically distributed, that integrates individuals' and partners' knowledge, by using ICT tools.

Virtualness Strategy. Strategy aimed at adding to the economic benefits arising from geographical proximity, and the advantages arising from organisational proximity, based on Internetworking technologies.

Wireless e-Mail. Applications that allow users to send and receive e-mail through mobile devices. These applications, together with workgroup platforms form a sub-category known as Wireless Workgroup applications.

Wireless Product Identification. Applications that give companies the ability to identify products or parts without physical handling.

Index